Claude Ghaoui, Mitu Jain, Vivek Bannore, Lakhmi C. Jain (Eds.)

Knowledge-Based Virtual Education

Studies in Fuzziness and Soft Computing, Volume 178

Editor-in-chief
Prof. Janusz Kacprzyk
Systems Research Institute
Polish Academy of Sciences
ul. Newelska 6
01-447 Warsaw
Poland
E-mail: kacprzyk@ibspan.waw.pl

Further volumes of this series can be found on our homepage: springeronline.com

Vol. 163. A. Ghosh, L.C. Jain (Eds.)
Evolutionary Computation in Data Mining, 2005
ISBN 3-540-22370-3

Vol. 164. M. Nikravesh, L.A. Zadeh, J. Kacprzyk (Eds.)
Soft Computing for Information Prodessing and Analysis, 2005
ISBN 3-540-22930-2

Vol. 165. A.F. Rocha, E. Massad, A. Pereira Jr.
The Brain: From Fuzzy Arithmetic to Quantum Computing, 2005
ISBN 3-540-21858-0

Vol. 166. W.E. Hart, N. Krasnogor, J.E. Smith (Eds.)
Recent Advances in Memetic Algorithms, 2005
ISBN 3-540-22904-3

Vol. 167. Y. Jin (Ed.)
Knowledge Incorporation in Evolutionary Computation, 2005
ISBN 3-540-22902-7

Vol. 168. Yap P. Tan, Kim H. Yap, Lipo Wang (Eds.)
Intelligent Multimedia Processing with Soft Computing, 2005
ISBN 3-540-22902-7

Vol. 169. C.R. Bector, Suresh Chandra
Fuzzy Mathematical Programming and Fuzzy Matrix Games, 2005
ISBN 3-540-23729-1

Vol. 170. Martin Pelikan
Hierarchical Bayesian Optimization Algorithm, 2005
ISBN 3-540-23774-7

Vol. 171. James J. Buckley
Simulating Fuzzy Systems, 2005
ISBN 3-540-24116-7

Vol. 172. Patricia Melin, Oscar Castillo
Hybrid Intelligent Systems for Pattern Recognition Using Soft Computing, 2005
ISBN 3-540-24121-3

Vol. 173. Bogdan Gabrys, Kauko Leiviskä, Jens Strackeljan (Eds.)
Do Smart Adaptive Systems Exist?, 2005
ISBN 3-540-24077-2

Vol. 174. Mircea Negoita, Daniel Neagu, Vasile Palade
Computational Intelligence: Engineering of Hybrid Systems, 2005
ISBN 3-540-23219-2

Vol. 175. Anna Maria Gil-Lafuente
Fuzzy Logic in Financial Analysis, 2005
ISBN 3-540-23213-3

Vol. 176. Udo Seiffert, Lakhmi C. Jain, Patric Schweizer (Eds.)
Bioinformatics Using Computational Intelligence Paradigms, 2005
ISBN 3-540-22901-9

Vol. 177. Lipo Wang (Ed.)
Support Vector Machines: Theory and Applications, 2005
ISBN 3-540-24388-7

Vol. 178. Claude Ghaoui, Mitu Jain, Vivek Bannore, Lakhmi C. Jain (Eds.)
Knowledge-Based Virtual Education, 2005
ISBN 3-540-25045-X

Claude Ghaoui
Mitu Jain
Vivek Bannore
Lakhmi C. Jain (Eds.)

Knowledge-Based Virtual Education

User-Centred Paradigms

Springer

Claude Ghaoui
Liverpool John Moores University
School of Computing
and Mathematical Sciences
Byrom Street
Liverpool, L3 3AF
United Kingdom

Vivek Bannore
Convergeys India Services Pvt. Ltd.
Cisco Systems Incorporated
Guragaon
Haryana
India

Mitu Jain
Vcustomer India Pvt. Ltd.
Netgear Incorporated
New Delhi
India

Lakhmi C. Jain
University of South Australia
School of Electrical & Info Engineering
Knowledge-Based Intelligent Engineering
Mawson Lakes Campus
Adelaide SA 5095
Australia

ISBN 978-3-642-06408-1 **e-ISBN 978-3-540-32389-1**

ISSN print edition: 1434-9922
ISSN electronic edition: 1860-0808

Springer is a part of Springer Science+Business Media
springeronline.com

Softcover reprint of the hardcover 1st edition 2005

Cover design: E. Kirchner, Springer Heidelberg

Foreword

Preparing Society for Virtual Learning in the 21st Century

Virtual learning plays an important role in providing academicians, educators and students alike, with advanced learning experiences. At the forefront of these current technologies are knowledge-based systems that assess the environment in which such learning will occur and are adaptive by nature to the individual needs of the user (Grossman et al., 2003). The extent that the learner will benefit from such technology will depend on the educational setting that this service is provided.

In a number of traditional school settings, the primary channel of knowledge is through the classroom teacher. Virtual learning then plays more the role of Supplemental Instruction (SI), helping the student further explore the material and/or review the concepts already covered in class (Taksa and Goldberg, 2004). Yet, even at this level of education, there are two alternative settings that would demand that the source of primary education be provided virtually: 1) in rural (Reed, 2004) or international (Couchman, 1999) settings where resources are not as plentiful and training for qualified educators is not readily available; 2) in home-schooling in the United States which is on a steady rise (Princiotta et al., 2004).

In conjunction with the 2000 Census, the U.S. Census Bureau conducted a Current Population Survey (CPS) of a sample population of families across America with regards to Home Computers and Internet Use. The results showed a marked increase from that of a similar 1997 survey. The data presented here is rounded off so that trends are more easily understood; for the precise numbers, consult the reference for detailed tables. Some highlights of this data include the following items: 90% of all elementary and secondary school children had access to a computer, although only 80% showed significant computer usage in school. At home, 51% of households reported owning a computer overall, but within economic groups, this statistic varied from 30%

to 90% depending on household income. While computer and Internet access at school was equal across economic and ethnic groups, there was approximately 30% less access to computers at home for minority population groups. (Newburger, 2001).

The relationship of the above census data to the development of knowledge-based virtual learning is straightforward. Governments have a moral and social obligation to ensure equal access of education to all of its groups of citizens. As this research is successfully applied to intelligent tutoring systems, governments will become more obligated to ensure equal access to computers and internet usage, both in school and at home, which will then broaden the impact of the research and help create a more educated and equal society.

Another situation in which there is a growing need for knowledge to be provided virtually is that of job training and career advancement (Schank, 1997). Many companies encourage their workers to learn advanced techniques that will improve their performance on the job; these companies will even accommodate employees' schedules to accomplish this, but do not necessarily have the funds to hire professional trainers and as such, rely on knowledge-based systems to perform these educational functions. However, as the economy gets tighter and skills required are specialized, management will "buy" (hire) ready-made labor from outside the company rather than "make" (educate) labor from within. (Meares and Sargent, Jr., 1999).

Efforts are being made to train workers in a virtual environment. Funded in part by the University of Texas, a virtual learning program EnterTech simulates a realistic work environment with virtual co-workers, a supervisor, and human resources (Dean, 2000). This course can be accessed online over the Internet or offline by using CD-ROMs. The program targets entry-level positions in industrial and technology-based settings and trains individuals for the proper handling of materials, efficient warehousing, and clerical tasks such as answering phones. The course has both individual and collaborative components to complement the training. An ambitious program that provides free virtual job training to all of its citizens is the Technical and Further Education (TAFE) system of Victoria, Australia. One of its courses aims to enable individuals to obtain the International Computer Driving Licence (ICDL), a computer literacy standard endorsed by the Australian Computer Society (Pace, 2001).

The educational scenerios presented here underscore the importance of the present volume of research, "Innovations in Knowledge-Based Virtual Education." This compendium under the editorship of L.C. Jain and C. Ghaoui brings together leading researchers in the field, addressing the plethora of issues involved in advanced learning technologies. Lakhmi Jain is an expert in knowledge-based systems and is the Co-Director Web Intelligence Consortium (WIC) Australia Centre and Professor of Knowledge-Based Engineering at the University of South Australia. Claude Ghaoui is a Senior Lecturer at the School of Computing and Mathematical Sciences, Liverpool John Moores University and has published and edited numerous works on e-education and is the editor of the Encyclopedia of Human Computer interaction (Information Science Publishing, USA; ISBN: 1-59140-562-9).

This work comes at a critical moment of educational development, as the world goes online and communication between all people is fostered at an ever-increasing rate. It is the hopes of this author that educators and leaders worldwide will utilize the technologies developed in this book to provide people with proper education and training.

Robert Goldberg
Professor
Queens College,
Flushing, NY 11367 USA

References

1. Couchman, J. A. (1999). Distance PALS in real and virtual classes. In Proceedings of the First National Conference on Supplemental Instruction and Video-based Supplemental Instruction (pp. 32–46). Kansas City, MO: Center for Supplemental Instruction, University of Missouri-Kansas City.
2. Dean, K. (2000). Virtual Training for Real Jobs. Wired News, March 22, 2000. Accessed from http://www.wired.com/news/culture/0,1284,33897,00.html.
3. Grossman, L. K., Minow, N. N. and Murphy A. G. (2003). Creating The Digital Opportunity Investment Trust (DOIT), A Proposal to Transform Learning and Training for the 21st Century. A Report to The Congress of The United States. Accessed from http://www.digitalpromise.org/about/report_to_congress/Executive_Summary.pdf. Detailed tables can be accessed from http://www.census.gov/population/www/socdemo/computer/ppl-175.html.
4. Meares, C. A. and Sargent, Jr. J. F. (1999). The Digital Work Force: Building Infotech Skills at the Speed of Innovation. US Department of Commerce, Office of Technology Policy, July 1999.
5. Newburger, E. (2001). Home Computers and Internet Use in the United States: August 2000. U.S. Census Bureau Report P23–207. Accessed from http://www.census.gov/prod/2001pubs/p23-207.pdf.
6. Pace, B. (2001). DEET/TAFE Virtual Campus. DMR Consulting Review, March 2001. Accessed from http://www.egov.vic.gov.au/pdfs/DEET_2.pdf.
7. Princiotta, D., Bielick, S. and Chapman, C. (2004). 1.1 Million Homeschooled Students in the United States in 2003. National Center for Education Statistics (NCES) Report #2004–115.
8. Reed, B. (2004). Providing Supplemental Services to Rural Schools. In NWREL Report: January-February 2004. Accessed from http://www.nwrel. org/nwreport/2004-01/Jan-Feb04.pdf.
9. Schank, Roger. (1997). Virtual Learning: A Revolutionary Approach to Building a Highly Skilled Workforce. New York, NY: McGraw-Hill.
10. Taksa, I. and Goldberg, R. R. (2004). Web-Delivered Supplemental Instruction: Dynamic Customizing Of Search Algorithms To Enhance Independent Learning For Developmental Mathematics Students. Mathematics and Computer Education Journal, Vol. 38(2), pp. 152–164.

Preface

The consideration of people with diverse needs and requirements must be taken seriously in order to provide innovative offerings in education. This can be achieved by employing solutions that are smart technologically and also sensitive to users'/human needs (i.e. user-centred). Widening access requires reducing disabling conditions under which users of education work.

Motivated by this challenge, the book provides various innovative approaches and principles that can be employed to further advance developments in educational technology, specially 'virtual education' (also known as online- or e-learning). In pursuing this motivation, the book promotes the continuous need to push for technology that serves people better. This requires innovative solutions that place users at the centre of concern and that can adapt to support many different groups of users; an issue, which unfortunately is still overlooked in most of the commercial systems, research and developments in this field.

The primary objective of this book is to provide a wide range of innovative approaches that can benefit various stakeholders (as users) of virtual education. In order for such approaches to succeed, the need to take multi-disciplinary and/or inter-disciplinary approaches is emphasized and followed by the authors. In doing so, the book fills-in a gap in this area, which is particularly invaluable to practitioners. The book is aimed at researchers and practitioners from academia, industry, and government, for an in-depth coverage of a broad range of issues, ideas and practical experiences on this subject. It also aims to raise more awareness in this important subject, promote good practice, share and evaluate experiences and lessons learnt.

This book includes 9 chapters. The following presents a brief overview of each chapter:

Chapter 1: Just-in-Time Approach to Learning: Arguing the Case for Cost-Effective Knowledge Dissemination, by M. A. Rentroia-Bonito, J. Tribolet, J. A. Jorge and C.Ghaoui. In order to provide e-learning that is timely and cost-effective, this chapter argues the need for taking a holistic approach from different perspectives: organisations, business, people and systems. It presents a possible framework to help achieve this goal, and assesses its usability for future improvements.

Chapter 2: P-Dinamet: A Web-Based Adaptive Learning System to Assist Learners and Teachers, by Miguel Montero1 and Elena Gaudioso2. This chapter presents P-Dinamet, a web-based educational system and the pedagogical model behind it, that can adapt to both teachers and learners, using knowledge acquired about these users. It describes the mechanism followed to make the system adaptable.

***Chapter 3**:* Intelligent Agents that Learn to Deliver Online Materials to Students Better: Agent Design, Simulation and Assumptions, by Leen-Kiat Soh, Todd Blank, and Lee Dee Miler. This chapter discusses an integrated framework of case-based learning (CBL) in an intelligent agent that can deliver learning material to students. The agent's reasoning is based on its learning of various cases/problems faced. A simulation that was built to test the agent's learning behaviour is discussed in the chapter.

***Chapter 4**:* Intelligent Web-Based Computer-Supported Collaborative Learning, by Vladan Devedzic. This chapter provides an overview of issues in computer-supported collaborative learning (CSCL), in relation to intelligent web-based learning and their evaluation. It addresses modern techniques for this purpose, e.g. web mining and intelligent agents.

***Chapter 5**:* Using Multiagent Intelligence to Support Synchronous and Asynchronous Learning, by Xuesong Zhang, Leen-Kiat Soh, Hong Jiang, and Xuli Liu. This chapter describes a system called I-MINDS, an innovative multi-agent system to support synchronous and asynchronous cooperative learning, both in a real classroom situation and in distance education. The chapter discusses different aspects of the system's design and evaluation, and gives some useful insights.

***Chapter 6**:* Intelligent Agents to Improve Adaptivity in a Web-Based Learning Environment., by C. I. Peña1, J. L. Marzo2, and J. Ll. de la Rosa2. This chapter focuses on the use of intelligent agents to support specially online life-long learners. It uses a multi-agent architecture called MASPLANG, which is adaptable, by its ability to build a hybrid student model, starting with knowledge about learning styles of the students, which then gradually modifies this knowledge when more learning about the students is acquired from his/her interactions.

***Chapter** 7:* Intelligent Virtual Teaching, by Goran Simic, Dragan Gasevic, Zoran Jeremic, Vladan Devedzic. This chapter discusses the main characteristics of intelligent learning management systems (ILMS) and identifies their shortcomings and advantages. It also describes as an example of such systems, a semantic web enabled system, called Multi-tutor, and discusses how using semantic web technology could benefit e-learning.

***Chapter 8**:* Developing a user-centred model for creating a virtual learning portfolio by M.Verhaart and Dr Kinshuk. This chapter discusses a personal content management framework, called 'Me', that gives individuals the ability to create a personal electronic portfolio of their knowledge, based on various sources, like: instruction, research, discussions, experience, insights, feedback, etc. The chapter describes two models (called 'Sniplet' and 'Multimedia Object'), upon which content in 'Me' is structured. It also overviews prototypes implemented to test this framework, and assesses the findings.

***Chapter 9**:* A Didactics Aware Approach to Knowledge Transfer in Web-Based Education, by Denis Helic, Hermann Maurer and Nick Scerbakov. This chapter presents an innovative system called 'WBT-Master' that was mainly created to support didactics aspects of web-based education (e.g. collaboration, project-oriented learning, or experiential learning) in order to facilitate transfer of knowledge among users. Useful insights are given and discussed.

The chapters included in this book cover a wide range of important issues on the subject of "*Innovations in Knowledge-Based Virtual Education* ", representing experiences from several countries. The chapters report on research, development and real experiences, including theory, practice, techniques, analysis, design and work in progress. Authors presented insights and views, by reflecting on the inter- and multi- disciplinary nature of this topic, addressing it from different perspectives. The main contribution of this book is in its focus on innovative solutions for the benefit of Virtual Education.

We are grateful to the authors and the reviewers for their valuable contributions. We appreciate the assistance of Feng-Hsing Wang during the evolution phase of this book.

Editors

Contents

Contents

1. Just-in-Time Approach to Learning: Arguing the Case for Cost-Effective Knowledge Dissemination

M. A. Rentroia-Bonito[1], J. Tribolet[1], J. A. Jorge[1] and C. Ghaoui[2]

[1] Departamento de Engenharia Informática, Instituto Superior Técnico, Av. Alves Redol, 9, Lisbon, Portugal

[2] School of Computing and Mathematical Sciences, Liverpool John Moores University, UK

This chapter focuses on e-learning and its relationship with organizational knowledge dissemination. We argue this requires a holistic approach that involves business processes, people and systems to deliver knowledge required by organizations and its workforce in a timely and cost-effective manner. Organizations need to be adaptable and flexible to stay competitive, which is a challenge they face everyday. Our holistic approach to online learning, as presented in this chapter, was motivated by this challenge. For this, organisations need to create and sustain an internal communication space to deal with a diverse and global workforce. Complexity is usually increased, as this global and diverse workforce typically performs process-based roles across different functions, priorities, contexts and cultures, whose content dynamically changes in accordance with business environment fluctuations. However, measuring profitability of this kind of initiatives is still an organizational challenge. This contribution represents a first step towards developing a theoretical framework to approach the creation of such a communication space as related to virtual learning and knowledge dissemination. To achieve this, we call for cost-effective organizational knowledge dissemination, as a main condition for just-in-time virtual learning. Furthermore, we raise some key issues to consider when developing proper measurement tools. We look at knowledge dissemination dynamics and their potential relationship to e-learning and its acceptance levels within organizations.

M.A. Rentroia-Bonito et al.: *Just-in-Time Approach to Learning: Arguing the Case for Cost-Effective Knowledge Dissemination*, StudFuzz **178**, 1–22 (2005)
www.springerlink.com

1.1 Introduction

Structuring organizations based on a functional-oriented approach to achieve business goals have created unclear or different interpretations for participants even when they share same job objectives, thus, affecting the pre-requisite conditions for communication and work-related performance. Communication, as a business process, plays a critical role in sharing operational knowledge and overcoming constraints put by "*experts' scalability problem*" to cost-effectively achieve business objectives. A current organizational challenge is searching for news ways to systematically sustain a common space for communication across organizational levels, locations, and culture and workforce skills. However, current organizational efforts dealing with internal knowledge dissemination are not delivering expected results [15]. At this regards, technology can assist organizations in facing this challenge and contribute cost-effectively to organizational learning by helping them: (a) create a structured conversation space, and (b) frame interactions between people and systems. So, how can we create and sustain this conversation structure in operational terms framing interaction between humans and systems?

In this chapter, we explored, within a defined theoretical framework, the organizational knowledge dissemination process associated to the concept of just-in-time learning and uniview. For the purpose of this chapter, just-in-time learning is strategic knowledge acquisition enmeshed in business activities to support employees in learning new skills when performing day-to-day tasks, while fostering the alignment between learning outcomes, technological and strategic business issues [23]. Regarding uniview concept, this concept forms itself as a consequence of the impact of group dynamics' stages on each member's perceptions, attitudes and behaviors. By building on literature [5, 20] and incorporating insights from Organizational Behaviors and Human Resources professionals, uniview is defined as a degree of shared understanding among team members on key issues of a problem, situation, event or solution. The formation of uniview ("*unique view*") strongly depends on internal communication and training initiatives [20]. Organizational efforts in operationalizing these two concepts, we hope, would help out organizations to face current challenges, particularly expert scalability and cost-effective skill development. To meet this particular challenge, next production and consumption of information and knowledge is addressed.

Since useful organizational knowledge capturing, updating and dissemination to the proper targets are a costly task which involves diverse roles and resources, and are a base for performance and continuously source for knowledge generation and competitiveness, understanding the systemic nature of knowledge production and dissemination is important to define proper strategies to align organizational resources to achieve expected results.

Thus, our proposal goes through structured organizational knowledge dissemination and is driven by an interaction paradigm based upon "*learning just what is priority for me to* ***do my work better*** *now*" instead of having access to all available information suffering consequently from unproductive overloading. This searched key differentiator factor is knowledge disseminated to whom needed it to improve performance. In this sense, our objective is to show relevant issues in cost-effective organizational knowledge dissemination and the theoretical framework behind it.

We hope, the main contribution of this chapter would be to provide a holistic framework towards understanding how the knowledge dissemination dynamics would work, its potential relationship to e-learning and setting up the basis to manage communication spaces within teams fostering consistency and its acceptance levels within organizational settings. This is a starting journey, whose increasing understanding would allow focusing efforts cost-effectively to help both side of the equation: producer and consumers of disseminated knowledge.

This chapter includes six sections. The first Section is an introduction. Section 2 (Cost-effective knowledge dissemination in organizations) summarizes main related concepts to knowledge dissemination. Section 3 (Cost-Benefit Analysis) describes key cost issues arising from each component of proposed conceptual framework. Section 4 describes our theoretical framework based upon reviewed literature. Finally, in sections 5 and 6, we present a general conclusion and future trends.

1.2 Cost-Effective Knowledge Dissemination in Organizations

To ensure knowledge dissemination, organizations should create a strategic context for sharing knowledge. According to Rosenberg [24], this strategic context requires as a key tool "*...a system for capturing, organizing, and storing knowledge and experiences of individual workers and groups within an organization and making it available to others in the organization*".

Effective implementation of this tool is four-fold. First, organizational policies and goals should enhance, facilitate and reward, expected knowledge sharing from recognized internal or external human experts. Second, investments in technological infra-structure should support the capture of the distributed operational knowledge from any organizational location and identified knowledge workers and guarantee the proper management for that structured knowledge. Third, a knowledge-management process should be in place to: (a) periodically monitor results, (b) timely identify/update pool of knowledge producers (those who transform data into knowledge), (c) continuously guarantee quality content in accordance with defined criteria, and (d) administer system's profiles and privileges. A last aspect relates to knowledge dissemination to other people in the organization.

We believe cost-effectiveness of organizational knowledge dissemination comes from given the right and proper amount of information to who needs it to perform their duties. This strategic context would contribute not only to consistency in organizational learning but also to bottom-line results savings costs in both side of the equation: (a) knowledge generation and dissemination, and (b) knowledge consumption or utilization. Basic assumptions at this point are the: (a) adequacy of interaction paradigm, whose main symptom is information overloading and eventually network underperformance, and (b) technology role as an intermediary between people and business processes. The former gives life to the just-in-time learning concept, the latter relates to the increasingly possibilities to make smarter systems to support users' tasks in achieving their goals in contexts of use pervasively. Given the existence of this strategic and technological context for knowledge sharing to take place, we next focus on the dissemination of that business knowledge.

By definition, knowledge dissemination intends to distribute knowledge to many people or organizations. As Figure 1 shows, knowledge as an organizational asset is costly to produce and maintain facing operational demands and constraints regarding highly uncertain business environments. Therefore, dissemination must be a cost-effective process not only technologically but also in addressing the proper target interested in using or consuming that knowledge. This means that knowledge should be disseminated among process-based roles needing it and motivated to use it. In this sense, wasting organizational resources would diminish in both side of the equation: organizations do not spend resources to distribute information to everyone and individuals do not waste time in dealing with information overloading.

To achieve cost-effectiveness, on one hand, each role should be analyzed to explicitly identify incumbents' needs, goals and required resources to perform tasks regarding knowledge usage patters. On the other hand, required skill levels must be developed and people should be motivated and committed to use or re-use knowledge repositories.

Two common mechanisms could be through communities, formal or informal, collaborative work and monitoring. Behavior imitation, social pressure, relevant supervisor support, usable and useful systems could help to reinforce motivation to use disseminated knowledge. This way, knowledge dissemination could be cost-effectively distributed minimizing information overload among those roles' incumbents that do not need sharing of knowledge, saving costs in both side of the equation: production [15] and learning at operational levels. Key indicators could help identify impact, and potential improvement areas, on bottom-line results of: (a) opportunity and direct cost of allocated organizational experts' time in production and reviewing, (b) technology-related costs to knowledge capture, dissemination and monitoring, and (c) effectiveness of current human resources policies and strategies regarding business strategies [7].

1.3 Cost-Benefit Analysis

Traditionally, training programs are assessed at four levels, as Kirkpatrick's model indicates: (a) trainee reaction, (b) knowledge transfer; (c) behavioral change and (d) business results. However, current general results from e-learning efforts and related start-up investments have made decision-makers put closer attention to cost-benefits analysis of such initiatives. Main objective of an e-learning cost-benefits analysis is to determine the economic feasibility of setting up an e-learning initiative to develop specific people competencies aligned with business strategies. Even though, this is a complex task, due to the nature of e-leaning, based upon reviewed literature [7, 16, 15, 20, 22, 24] and insights from IT and

Human Resources professionals, Table 1 summarizes some aspects related to organization, technology, physical infrastructure and people and expected benefits in setting up organizational e-learning initiative taking into consideration organizational and technological layers, as shown in Figure 1.

Table 1. Associated costs and expected benefits

Organization-related	
1. Cost of recruiting new employees	Less turnover rate
2. Opportunity cots of experts or seniors during learning content production, repacking materials, reviewing process and lecturing	(a) Maximizing work-related knowledge dissemination effort and consequently impact on reutilization of internal organizational operational knowledge; and (b) Strategic areas of performance, cultural values, operational methods and processes, among others, consistently disseminated (asynchronous and synchronous) by process-based roles contributing to a sustained organizational space for internal communication
3. Impact on team productivity level during initial learning phase (ex. work continuity, motivation issues, etc.)	(a) Easier time for new team members in caching up group dynamics and performance levels; (b) Best performers are trained too, and (c) Loss of individual productivity during learning
4. Opportunity cost of trainees	(a) Individual performance improvement; (b) Speeding up learning curve; (c) Faster adaptation to job requirements, and (d) Improved efficacy of elearning programs
5. Project management costs	Warranting delivery of agreed outputs on time and up to quality levels during e learning program development cycle

Technology-related	
1. Equipment for Client/Server architecture involving key parts (instructors, authors or content producers and reviewers, transaction costs, etc.)	IT investment aligned with Business strategies
2. Development and maintenance of e-learning platform	(a) High business strategy, process and system fit; (b) definition and implementation of contingency plans
3. System administration and Helpdesk staff	(a) Operational efficiency and flexibility; (b) Change management, and (c) Monitorization of system performance and utilization patterns
4. Other equipments (ex. videoconference, air conditioning, reflectors, microphones, webcams, etc.)	(a) Standardize online training process, and (b) Continuous reutilization and fine-tuning according to e-learning program' results
5. Telecommunication costs	(a) Low costs, and (b) increased bandwidth and network performance

Physical infrastructure -related	
1. Setting up local and remote physical settings	(a) Standardize technical conditions across involved partners, and (b) Development of adequate options in accordance with current, and evolving, system performance
2. Facilities and Maintenance costs	known fixed costs

Training process-related	
1. Hand-outs and others printed materials (ex. lectures scripts, syllabus, questionnaires, etc.)	Standardize methods of material and content delivery to manage adequately people's expectations
2. Orientation sessions for potential e-learners, Marketing events (online and offline) and promotional materials (ex. pamphlets)	(a) Better management of people's expectations, and (b) identify specific needs to drive re-packing efforts.
3. Quizzes and exams	(a) Flexibility (covering local and remote learners), (b) Adjustment to institutional requirements, (c) Timely feedback; (d) Change management
4. Assistant staff (ex. updating online material, publishing training related announcements, readings and results, moderating forum, etc.)	Timely learning content management
5. HR staff (ex. Identifying potential short- and long-term organizational key competencies; assuring learning outcomes aligned with strategies and IT architecture; managing organization's Human Resources' training and development processes)	Better people-system-context fit to improve performance
6. Internal trainers(ex. Identifying potential training needs within organizational context for each process-based roles, defining proper instructional methods to target's specificities and respective media, preparing and delivering e-lectures; re-packing learning modules according to learning objectives and potential audience; monitoring e-learning programs' results and coaching and supporting e-learners, among others)	Adequacy among learning needs, methods and media taking into account the specificities of user group in each process-based role and type of learning content
7. Training e-learning staff (Assistant, Administration system, Instructors or internal trainers, administrative staff, Help-desk staff, etc.)	Framing change management process

People-related	
Individual resources invested in work-related priorities and goals according to personal pace of learning and convenience	(a) Learn as required by work-related activities and demands; (b) More effective transference of acquired knowledge; (c) Likely certification of acquired skills by educational institutions, and (d) improved employability

The main purpose of this summary is to contribute in analyzing main cost drivers associated to e-learning initiative taking into consideration the complexity and dimension of the e-learning initiative. This analysis includes operational costs required to establish operations, train staff, and implement appropriate information technology support when organizations decide to go from traditional to online learning environments, and also, involved some intangible costs, such as experts' and potential e-learners' opportunity costs (e.g., cost of experts while being involved in e-learning activities, lost revenues or increased costs or delays in delivering services to clients during participation in an learning session) and potential impact on individual productivity and satisfaction.

To calculate the investment required to implement an e-learning environment and the savings resulting from changing from current to expected situation, implementation costs should be estimated taking into account the previously-mentioned aspects, differentiating fixed and variable costs in terms of e-learning strategic choice. Thus, a break-even point could be defined in terms of defined strategies to reach out process-based roles' incumbents' specific needs and characteristics and disseminate organizational knowledge assertively.

Thus, the profitability of an organizational e-learning program will depend on its initial scope, synergies and expected development taking into account improvement areas identified in accordance with: (a) e-learners' real needs, performance, satisfaction, and acceptance, and (b) business strategies and choices. These issues would justify investment on e-learning initiatives. Fully operational e-learning environment would allow gradual cost savings around 50%-70% to overset initial investments during 3-5 years with a consistent group of actively engaged online leamers [16, 22].

For instance, initial phase might be to deliver, through Intranet, structured organizational content to specific group of learners to test the acceptance of learning online within organizational setting, or it maybe acquire an e-learning platform to manage content and training processes. In both case, organizational resources and readiness to embrace e-learning initiatives are key to define financial return. In order to frame theoretically the profitability analysis of an e-learning initiative, next section focuses on theoretical framework emphasizing the business strategy, technology, process and people fit as a condition to effective organizational e-learning experiences.

1.4 Theoretical Framework

Performance alignment is a major issue for organizational contexts to stay competitive. Productive communication, within increasingly borderless organizational settings, becomes a key issue for this alignment to happen, specifically optimizing resources to approach market-driven events. In particular, information resources as corporate asset are a relevant component to competitively act and perform. In order to understand the dynamic of knowledge production and consumption within organizational setting, we propose a theoretical just-in-time approach to learning addressing context-specifics and process-related aspects as well as the need to foster communication spaces that promote workgroup and expected performance level in solving complex business challenges.

The creation of a meaningful space for communication has become even more difficult due to the multidisciplinary approach demanded by increasing: (a) cross-cultural task complexity at tactical-strategic levels, (b) task simplicity at operational level, (c) existence of legacy systems supporting core activities or processes, (d) scarcity of human experts in critical operations, and (e) increasing difficulty in creating effective communication mechanisms that can make possible, rapid and cost-effective sharing or dissemination of operational business knowledge across organizational contexts, groups, units, levels or roles. Incomplete or poorly alignment between communication process, business goals and the proper target affects organizational performance and organizational climate in two ways. First, what individual perceive from their work environment impact their related behaviors and intentions to adopt new paradigms in the workplace [8, 14, 20].

Systems, being instrumental part of organizational settings as working tools and communication actors, may speed misunderstandings and confusion causing enormous damages to individuals and organizations if not ethically and securely managed affecting bottom-line results and related productivity levels. Second, development team's interpretation of users' reality and skills to specify system requirements is key to increase acceptance and usage [14]. In creating this expected communication space, it is required to define and promote a shared view on task, problem, solution or situation domain by people in order to synchronize expected business actions. In the following subsection, context-specific aspects of our theoretical framework are presented to support our approach in addressing these previously-mentioned challenges.

1.4.1 Context-Specific Aspects

Figure 1 shows a high-level view of our theoretical approach to learning when people perform business enmeshed into their roles. Context-specific aspects are represented as Organization and Technology layers, at macro and micro-organizational levels.

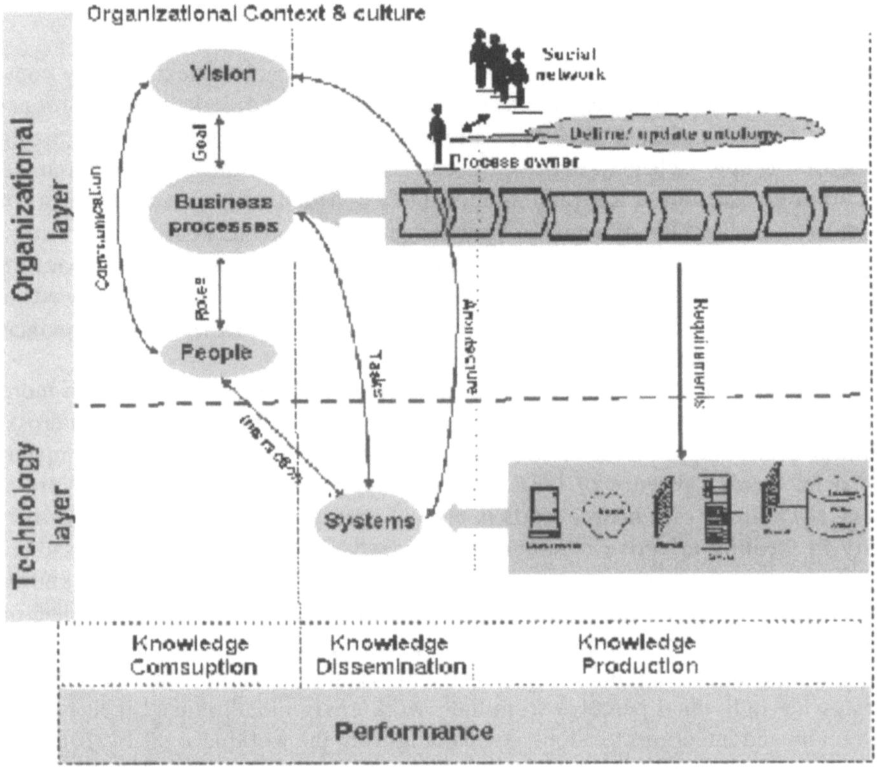

Figure 1: High-level view of our theoretical framework.

At macro-organizational level, vision drives architectural choices, process' goals, skill gaps and internal communication contents. Their articulation is expressed by an internal and external fit, which should be highly present to progressively establish a strategic context to synchronize processes' objectives, resources, goals, processes' main stakeholders and expected results. If organizations want high technology acceptance and usage, they have to harmonize work, skill and system development processes in order to create the necessary conditions to frame interactions between involved actors (humans or systems). At micro-organizational level, people and systems and its relationships (roles, requisites and tasks) define the required knowledge space to drive expected business performance levels. For the purpose of the remaining of this chapter, this is a so-called people-system fit.

Business processes act as mediators between vision and expected interactions at micro-organizational level, namely at immediate individual work context where individual and team performance take place. Business processes provide the business knowledge structure based upon process-related ontology which is represented by an internal fit. This fit means synchronization among all business processes within same work setting articulated with related business strategy. This fit:

(a) guarantees harmonization of efforts among processes' goals and required key skills at organizational tactical and operational levels, and (b) addresses the internal communication efforts to effectively manage different channels (offline and online) used to communicate with organization's main stakeholders which gradually build mental models on expected individual, teams' and organizational contribution levels. That is, organization and coordination of work is made around strategic choices to get expected results at process-based roles aiming at structural performance alignment.

As seen in Figure 1, the previously-mentioned context-specifics aspects clearly influence the quality and quantity of interactions between people and systems through, for instance: (a) usability goals driving a major consideration to human factors when developing information systems, (b) Human Resources policies supporting the development of competencies through expert's informal network by encouraging, and rewarding, knowledge sharing and coaching programs, (c) IT investments being adequate to organizational vision and current situation, (d) consistent internal communication initiatives promoting alignment between individual and organizational performance among others, and (e) methodologies and tools aiming at higher maturity levels in software development. Two main issues arise at this point: quality of service and quality of relationships.

Quality of Service relates to technology layer. In this layer, systems reflect technology architectural choices, supported tasks, interaction types and patterns. Methodologically speaking, this layer comprises the different views within Zachman framework when developing information systems [10]. Levels of quality of service come out from value chain's operational efficiency, specifically due to processes optimization or re-design and issues of network performance, bandwidth and the like. Quality of relationship regards to social and cultural aspects of work settings, expressed by leadership styles, coaching and mentoring programs, helpdesk mechanisms, supporting learning or interest community development, values (e.g. equity, respect, ethical behaviors, etc.), effectiveness in internal communication, and the like.

At micro-organizational level, this people-system fit translates into: (a) the development of a shared understanding among team members on key issues of a problem, situation, event or solution framed by business process' knowledge domain, and (b) expected business performance setting up a conversation structure among key actors (humans or systems). This fit results from the degree of adequacy between system requisites and users' and tasks' needs dynamically affecting its acceptance and usage at operational levels [8, 13, 19, 20].

Having briefly described relevant context-related aspects to just-in-time learning approach, next definitions, for the purpose of this work, of uniview and conversation structure are discussed before digging into how operational knowledge could be created and updated.

Building on theories from several disciplines (e.g. Human Resource Management, Organizational Behavior, Human-Computer Interaction, Learning and Social Psychology), we define uniview as a degree of understanding among team or organizations' view on key issues of a problem, situation, event or solution. It is a very operational term. This means a shared, and current, understanding of a group of people about a specific situation that demands an action (performance). Its development follows work group's dynamics and strongly influences interaction patterns among group's members. Therefore, its lifecycle is also similar: very light at the beginning of the interaction based just on individual background, skills, roles and points of view and varying over time according to the quality of interactions during workgroup lifetime [20]. As seen in Table 2, a critical second way, in which organizational performance could be affected, refers to competence development as a base to build uniview.

Table 2. Uniview stages and their relationships with Business processes, group Dynamics, Operational Knowledge and Potential Stakeholders.

Business Processes	Group development stages	Uniview stage	Operational knowledge type of activities	Potential stakeholders
Monitoring, Lessons learnt, Optimization.	Adjourning	Dissemination	Consulting	Pre-defined process-based roles' incumbents. Communities of Practice, Interest or Learning. Workgroups
Data capturing and analysis	Performing	Sharing	Updating	Team members and key decision-makers
Norms, methodology and tools, expected behaviors and results, individual objectives and responsibilities and training	Norming	Knowledge structuration	Creating/ reviewing	Process owners and related expert network
Defining work structure and roles. Identification training needs	Storming	Task-related language for group communication	Ontology, meta-data	
First meeting, bounding, coaching and Orientation sessions	Forming			
Process objectives, scope, available resources and roles/required competency				

Background sharing (e.g. Previous work experience, interests, work commitments and demands, group context and other context-related specifics) defines the primary space for common understanding among actors. At this point, process-related language, meanings and expected actions are different as so are interpretations of same words or requirements across involved actors. When basic language, meaning and expected actions are understood by all parties involved, a communication space is created where progressively uniview on a specific topic would be developed. To effectively build this communication space, people should be informed and trained, or coached, to make easy engagements into the defined conversation structure. To effectively build this communication space, people should be informed and trained, or coached, to make easy engagements into the defined conversation structure. This communication space should be based on a defined and shared ontology using high-usable systems.

This communication space should be based on a defined and shared ontology using high-usable systems. Experts' expected role is defining the process-related ontology within organizational context, and eventually acting as coaches in key process' competency areas. The process' owners would update defined ontology and reinforce a team-based approach to problem-solving or decision-making processes. This reinforcement would support the emergence of the communication space consolidating uniview on specific topics, and positively influence system's acceptance and later usage. This would support expected business actions/performance on specific situations.

The existence of such structured conversation protocol would minimize potential sources of misunderstandings among involved parties. This could contribute to the definition and reinforcement, as organizational values, of expected behaviors [20]. On the other hand, systems will support process-related knowledge distribution, and learning (defined by a network of specific acts or commitments, for instance communities of practice, forums, work groups, distance learning, etc.). This would progressively reinforce organizational discipline in dealing with these information resources [8, 13, 19, 24, 26].

This is the vision behind our proposed just-in-time approach to learning. In achieving this vision, main challenges are to set up the required conditions to build these communication spaces, timely develop required competencies and design usable, actable and intelligent systems in order to help people learn and continuously improve their performance when executing their roles. Next, three basic aspects for this framework are briefly described: (1) how communicative acts turn into performance acts , (2) how e-learning could develop required competencies, and (3) how usable and intelligent systems should be developed.

1.4.2 From Communicative to Performance Acts

Communication is defined as a two-way conversation, normally among people (speakers and listeners) in a flexible and unstructured way reaching sometimes diverse and unrelated knowledge domains and diverse emotional states as conversations evolved [26]. Communication acts reflect speaker's intentions to act and may influence listeners' beliefs, attitudes and also intentions to behave or perform in a dynamic way, especially when speakers are relevant to listeners [5]. Communication cycle stops when one of the involved parts leaves, gives up or "*disconnects*". Basic differences between face-to-face and online communication types are summarized into three categories: (a) high need of structuring conversational space; (b) low options to express emotions, (c) degree of credibility on communication source, and (d) confidence in systems as mediating instruments [8, 20,26].

Both communication types (face-to-face and online) share context-specificity as a commonality to gain meaning and be interpreted. Also, they must be dynamically consistent to minimize cognitive dissonances that may impair performance or expected work-related behaviors [20].

Consistency among communication types and channels is important for creating/sustaining the emergence of a communication space positively influencing system's acceptance and later usage supporting expected business actions/performance on specific situations [3, 14]. This communication space is defined by business words, or linguistic acts, related to business process' dynamics in context. This makes systems be communicator actors [4].

Linguistic acts generate different kinds of commitments depending on the participants' roles and involvement within the network of recurring conversations to which they belong. Speech Act theory [8, 26}, as shown in Table 3, defined three types of communicative acts.

Table 3. Types of communicative acts.

Locutionary acts	... relate to speaker's intentions and understandings of knowledge domain expressed through used words and prepositional content
Illocutionary acts	... relate to simultaneously actions done with locutionary acts (e.g. Promise, Acknowledging, Questioning, Accepting, Rejecting)
Perlocutionary acts	... are the communication effects (e.g. Response to received messages, changing knowledge, attitudes level in listeners).

One of main challenges in interaction is to set up a common understanding about a specific knowledge domain, whose sharing could serve as a basis for further actions (e.g. Knowledge acquisition and re-use, decision-making, data or information interpretations, etc.) within an ever-increasing interdependent work context, as illustrated in Figure 2.

This sharing is context-related and shapes individual interpretations, which are influenced by individual information gaps. If alignment between communication process (online and face-to-face), business goals and the proper target (e.g. role' incumbents) is incomplete or inadequate, it would affect organizational performance in two ways.

First, people interacting with a specific system normally do it because systems are expected to be useful in achieving their work objectives [8, 21]. To do so, people have to understand, not only be aware, but comprehend, from a specific point of view, the basis of the process-knowledge domain, its objectives, its context, its stakeholders, involved resources (where systems are categorized), the meanings of the related business words and the expected business actions that should be performed, by them or through systems. This understanding is sustained by continuous interactions overtime, with content, experts, team members, process owners or other sources such as social network [6].

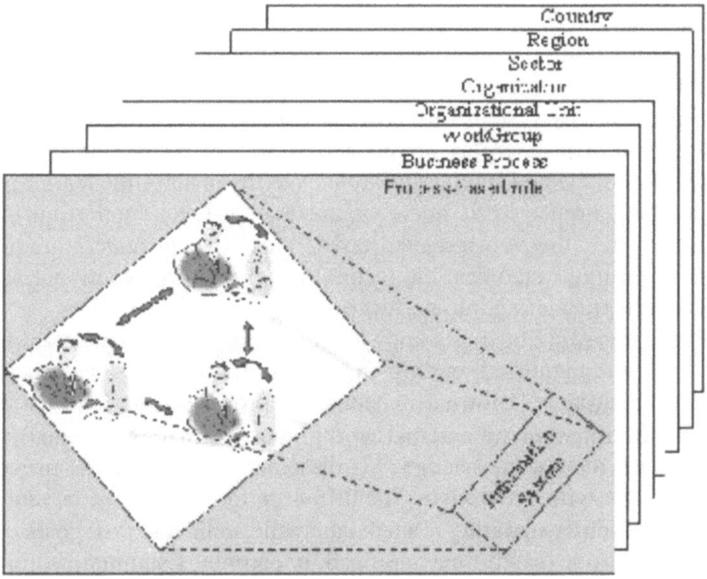

Figure 2: People-System fit in Context.

However, organizations normally have "*expert scalability problems*" and because of which continuous interactions are difficult to implement. Hence, having people with a "*unique view*" on the process domain without continuous interactions with expert sources could be a competitive advantage. Systems can help to meet this challenge, as previously mentioned. They should be: (a) designed to capture structured data from operational key processes, (b) define a conversation structure to support online communication, and (c) reinforce, through words, meanings and related business actions the expected contribution levels.

Second, systems, if properly used as organizational change agents, could contribute to improve people's perceptions on organizational procedural justice when guaranteeing process' transparency and equitable accessibility. This has an impact on organizational climate and effectiveness of standard operative procedures.

In short, key issues for influencing the relationship between communicative and performance acts are: (a) consistent conversation structure, (b) updated process-related ontology, (d) quality of content, and (b) teamwork approach to problem-solving or decision-making processes. The next subsection describes how the previously mentioned aspects relates to systems within the business process-people fit

1.4.3 Business Process-People Fit: E-Learning and Knowledge Dissemination Link?

As shown in Figure 1, Business Process, as structural factor, influences interactions between people and systems through process-based roles and system requisites. Process-based roles describe expected work-related behaviors and associated set of responsibilities for individuals to perform within business settings. System requisites are derived from task analysis and expected functionalities on key process issues. Interaction is: (a) defined by the relationships among roles within organizational business processes, (b) framed by formal and informal social network, (c) influenced by system's usability issues and (d) internal communication's consistency level. The interplay among these three relationships (Process-based role, System Requisite and Interaction) relates to the just-in-time learning concept: strategic learning immersed at micro-organizational level supporting individual learning when executing work-related tasks. This concept fosters gradual alignment among learning outcomes, the technological and organizational aspects of business contexts translating into performance alignment [23].

In addition, regarding business knowledge creation and distribution, four main characteristics are identified in this theoretical framework: (1) process-driven creation and updating, (2) formal existence of process-owner and its structured relationship with internal and external workgroups (formal or informal) with context-specific task-related knowledge, (3) dissemination based on process-based roles' incumbents with previously identified or forecasted needs, and (4) task execution continuously updating related data while achieving task goals.

The essence of our just-in-time approach to learning is summarized in one side in the interplay among role, system requisite and interaction, and on the other side, in businesses actions performed at operational level as expected at strategic levels, by: (a) investing in adequate IT; (b) effective internal communication strategy and process, (c) assertively developing required competencies, and (d) building progressively mental models regarding desired situations. Specific ways of aligning skill and system development are: (a) monitoring and reinforcing organizational internal fit, and (b) consistently implementing a participatory-type system development methodology.

Having briefly presented our theoretical framework, next subsection focuses on process-related ontology and its relationship to create and update operational knowledge.

1.4.4 Process-Related Ontology

As seen in Figure 1, data is created and updated when people perform tasks within their process-based roles. This way, operational data is captured by systems and feedback process. Process manager, by analyzing this information, would progressively: (a) update existing ontology, (b) create a high-quality and credible space for communication with users, (c) promote business-oriented actions with process-related words across contexts and workgroups, and (d) manage process-related information and related context-specific work-related knowledge. On one hand, users would receive related words through the content of their roles, which is process-based and highly supported by systems. Required competencies would identify skill gaps. This meeting point, between what the system "*knows*" about the user tasks and what the user "*knows*" about the system when performing tasks, is two-looped.

First loop relates to the process ontology. Ontology gains special relevance to make easy and usable knowledge structures to support later re-utilization. As shown, process owner defines ontology, which is business-driven. Updating ontology may come from inclusion of others' perspective (from members' social network) within the work process' domain. If properly managed, these new information sources may relate to process improvement or innovations [6].

Second loop relates to the definition of process-based roles' content. This resulting knowledge structure goes into the definition of roles. Roles are key for knowledge dissemination making straightforward the relationship between "*what is required to be known by those who need it to better perform their duties*". Incumbents are informed and trained to acquire/update skills. The reutilization of those knowledge structures and the flow of communicative acts, and consequent actions, across actors and process-based roles would then gradually be consistent and transparent, because it is structurally reinforced by systems and business processes. For instance, this knowledge structure can input the implementation of e-learning strategies as a way to improve skill development. If this is the case, updating these knowledge structures could be made from any location by process experts (authors or reviewers) in a cost-effective manner.

An expected result is two-looped flow is that the interaction between people performing their roles in interaction with systems would reinforce the development of the uniview concept, which is becoming increasingly important for organizations facing a global and diverse workforce and skill requirements. Before digging into developmental issues, next usability and actability aspects are briefly described.

1.4.5 People-System Fit

Technological advances let designers to increasingly "*mimic*" human acts (e.g. talk) representing reality into a linguistic structure of objects, properties and events (e.g. a person performs a process-based role within a specific organization and geographical location during a specific period of time interacting with designed systems in a multimodal way).

At this point, we need to be aware that systems could be tools for work, communication and action [3, 8, 9]. Two major concerns for system development are: usability and actability. Usability is the extent to which a computer system can be used to achieve specified goals with effectiveness, efficacy and satisfaction in a given context to complete a well-defined task [8]. Usability features gain special relevance to develop unique views among team members on specific situations cost-effectively. For instance, as shown in Figure 1, users engaged in interactions with systems if they perceive systems to be usable in achieving their role's objectives.

Information systems represent a structured knowledge domain where objects, properties and actions are designed accordingly to requisites. Their acceptance and usage level will vary upon system usability levels (easy-to-learn and use, useful for the task at hand, accessible to them) and also system's perceived quality of the data. Interaction exists, given connectivity and accessibility to data, if systems support users' tasks and communication flows among them. Users' tasks, commitments and also learning needs are defined within process-based role. This may imply, given all the possible events coming from user's context at any level, that a definition of a conversation structure within context is a must to standardize business responses or actions across individuals engaged in the same communication space [26]. This enhances the importance of uniview concept within the domain related to that process-based role. At this point, actability definition is in order.

Actability is mainly a performance of action concept supporting interactive actions within a specific context. These actions are instances of explicit intentions of generic business actions structured into business processes following generic business patterns [3]. This process structure supports the strategic alignment of the Organizational layer, shown in Figure 1. Also, they represent "*...knowledge about action intended for action.*" and are oriented to the behavior of others, which is usually purposeful and implicit [3]. Hence, an actable system is a system able to perform actions, permit, promote and facilitate the performance of actions by users, both through the system and based on information from the system, in some business context [4], is a high-usable system that could collaborate and act, if properly used, as a team member [21] to achieve workgroup's goals.

Actability comprises two types of knowledge that the system should have about the hierarchy of: (a) business actions starting in patterns and (b) individual's actions starting in user groups [3, 11, 17, 18, 24]. The goodness of fitness between these two hierarchies would allow two-way communication. This common space for communication is sustained by the existing level of: (a) shared meaning on the defined ontology or taxonomies of information and the existing rules of business processes and (b) trust among involved process-based roles' incumbents, and (c) existing mechanisms for communication offline and online. Next, some ideas on how can development teams gain the required understanding from work dynamics within roles.

Indeed, as shown in Figure 1, this conversation structure is defined as a relationship among business processes, systems and people defining the context for communication (messages or meanings) and performance (actions) at micro-organizational level. This context for communication changes over time because of: (a) systems' upgrades or technological architecture changes; (b) development of competence levels, (c) optimization of processes or tasks. Thus, systems should be usable and also can be actable in order to get a high people-system fit. Next, system development methodology is briefly addressed regarding its role in our approach.

1.4.6 System Development Methodology

Around 70% of large complex software projects fail to deliver expected results [13, 27]. Some development efforts fail to produce anything at all or produce products that people do not like and consequently will not use at all or if they do, they may use reluctantly and ineffectively. System development methodology and tools have evolved to better support system development. However, a more business and user-centered approach to technology development is required in establishing a systematic loop between user's needs and knowledge transference into the design of systems. This kind of forward-backward loop between users and development teams influence key dimensions of system's functioning and performance. Outcomes of this forward-backward loop would best match system performance to users' tasks, goals and needs, gradually improving the adequacy of defined system requisites [8, 10, 13, 14, 19, 27].

Towards this end, two relevant aspects are discussed in this subsection. First, building systems specifications should be process-driven reflecting its language as a basis to: (a) map current situation taking into account organizational and social aspects, and (b) identify related gaps. On the other hand, required skill levels to perform through the system within their process-based role should be properly and timely developed in order to improve acceptance and usage levels.

And second, within a work setting structured functionally, understanding the usefulness of systems for their intended users and anticipating their potential breakdowns, is not an easy task. Development teams should include system developers, process owners and user representatives in order to aim at improving people-system fit. That way, designed systems will help frame or update individuals' knowledge maps on a specific situation if interaction is properly adapted to user target.

However, this kind of fit requires a definition of structured knowledge domains where interaction would take place based on common pre-understanding with a minimum of words and conscious effort [26], namely a shared conversation structure. For this to happen, development team must understand user's language to: (a) represent how people understand and recognize patterns within their immediate work context's realities, and (b) design systems by interacting with users from the very beginning of the development cycle in order to contribute to a more effective, and less frustrating interaction. A solid understanding of the conversation structure by development teams would help translate user's language into requisites, allowing them to: (a) represent how people understand and recognize patterns within their immediate work context's realities, and (b) design systems by iterating with users since the very beginning of the development cycle. This participatory design approach will provide design with a more effective and less frustrating interaction.

For instance, development team's members, after interacting, gradually acquired uniview on problem and technology domains in accordance with their roles and respective objectives and interaction patterns.

1.5 Conclusions

The main contribution of this chapter was to provide a holistic framework to better understand the knowledge dissemination dynamics, their potential relationship to e-learning and its acceptance levels within organizational settings. As shown in our knowledge dissemination case, cost-effective knowledge dissemination anchor into the dynamic of business processes assuring consistency and minimizing waste of organizational resources. Work-related learning reviewed or produced by internal experts, available at process-based role and periodically updated, could aim at organizational learning in a cost-effective manner, though the initial high investment, and progressively becomes an organizational competitive advantage.

Even though, holistic approaches do not fully capture the whole dynamic and complexities of the organizational learning phenomenon at work settings, we hope to have contributed to identify some aspects of a business-oriented framework to approach the operational knowledge production, dissemination and consumption stages.

1.6 Future Trends

Four main trends are identified. First, learning services are going to be experienced as work-related events (e.g. problem-based scenarios, case studies, projects, etc.) taking place within process-based roles in just-in-time modes and diverse modalities. As technology becomes increasingly pervasive and intelligent, instructional design, ethical, and security issues of using e-learning systems, such as personal assistant, coaches [1, 2, 25] or representatives or communication actors, are in order for further research. If something goes wrong, who is responsible for this technological persuasion?

For example, within learning environments, courseware, as a social and mediating tool between instructors and learners, could facilitate, within the boundaries of the structured conversation space, recognition of work problems and other context-specifics aspects, and closely relate to learners' goals and expectations. A potential effect of interaction will follow (e.g. issue awareness, changes in knowledge levels, specific work-related actions), thus fostering specific business actions and its quality levels. Thus, the governance model of e-learning should be carefully analyzed taking into consideration these still evolving open issues.

Second, and as a matter of reflection, we think that the higher uniview within the project team exists, the closest the system specifications to user's language and reality would be (given the proper approach and managerial support). A high process-people-system fit could mean consistency among internal flow of communication and resources at organizational macro- and micro-level, and a team-based approach to problem-solving and decision-making within problem's or solution's knowledge specific domains. Project team's dynamics within context should be monitored in order to identify effective team behaviors and analyze its relationship with project' outcomes during project lifecycle in order to develop required competencies, methodologies and tools to assure higher acceptance and success rate.

Third, based on the theoretical proposition, that consistency in messages (online and offline) matters for performance, we believe that its level among messages across communication channels (humans and systems) should reinforce each other to minimize cognitive dissonance and consequently enhance motivation and productivity levels. In this way, business processes and systems fit would translate into correlations between business and individual role-related actions. In this sense, literally "*words have meaning*" that makes sense within a specific context for a particular user's profile. Managing properly this communication aspect –

consistency - would act as an "*attraction*" anchor for people sharing similar profile (role position, personal and professional interests on defined ontology [12], objectives, background, confidence level in communication sources [5, 6], among others). The effectiveness of this type of anchor should be further explored within the context of virtual communities.

To end, we believe that physical social network influences organizational members' active participation and acceptance of work-related systems. At this point, trust is a key driver. In what conditions, do people trust systems as work-related communication actors? Perhaps, ethnographic studies are needed to better understand how well users accept actions made by systems and integrate them within their context of daily work. This way, the ever-present vision of HCI of making interactions between humans and computers positively and wholly humane would progressively be enmeshed into IS tools assuring users during their tasks' performance.

References

1. Angehrn A.; Nabeth, T. and Roda C. (2001): Towards personalized, socially and active e-learning systems (illustrated with the agent-based system K-InCA. Centre for Advanced Learning Technologies, INSEAD. Retrieved on July 2003 from htpp://www.insead.fr/CALT/K-InCA-whitepaper.pdf.
2. Angehm, A.; Nabeth, T.; Razmerita, L. and Roda C.: K-InCA: Using Artificial Agents for Helping People to Learn New Behaviors. Proceedings IEEE International Conference on Advanced Learning Technologies, IEEE Computer Society, Madison USA, (ICALT 2001) 225-226. Retrieved on July 2003 from htpp:// www.ac.aup.fr/~roda/publications/PagKincaCALTpaperFinal.pdf
3. Agerfalk, P.J. (2001): Researching the applicability of Actability – Towards an Improved Understanding of Information Systems as Tools for Business Action and Communication. Proceedings Conference for the Promotion of Research in IT, Sweden, Part I: Scientific Contributions, Ronneby, Sweden, April 23-25.
4. Agerfalk, P., Sjostrom, J., Eliason, E., Cronholm, S. and Goldkuhl, G (2002): Setting the Scene for Actability Evaluation – Understanding Information Systems in Context. 9th. Conference on IT Evaluation, Paris, July 15-16 (ECITE).
5. Bandura, A. (2000): Cultivate Self-Efficacy for Personal and Organizational effectiveness. In E.A. Locke (Ed.), Handbook of Principles of Organizational Behavior. Oxford, UK: Blackwell. 120-136.
6. Conceição, P. and Heitor, M.: Systems of innovation and competence building across diversity: Learning from the Portuguese path in the European context. In Larisa V. Shavinina (Ed.). International Handbook on Innovation, Elsevier Science (2003). Retrieved on July 2003 from http://www.bolseiros.org/conferencia/Portugal Chapter.pdf.
7. Devaraj, S. and Ramesh, S. (2004): How to measure the relationship between training and job performance. Communication of the ACM, Vol. 47, N° 5 63-67.
8. Dix, A., Finlay, F, Abowd, G. and Beale, R. (1998) Human Computer Interaction. Prentice Hall Europe, Second edition.
9. Ehrlich, D. (2002): Establishing Connections: Interactivity Factors for a Distance Education Course. Educational Technology & Society 5 (1), 48-54.

10. Inmon, W.H., Zachman, J. and Geiger, J. (1997): Data Stores, Data Warehousing and the Zachman Framework, Managing Enterprise Knowledge. McGraw-Hill.
11. Hara, N. and Kling, R. (2000): Students Distress with a Web-based Distance Education Course: An Ethnographic Study of Participants' Experiences. Information, Communication & Society 3(4) 557-579.
12. Kalfoglou, Y.; Alani, H.; O'Hara, K. and Shadbolt, N.: Initiating Organizational Memories using Ontology Network Analysis. Knowledge Management and Organizational Memories workshop, 15th European Conference on Artificial Intelligence, Lyon, France (2002). Retrieved on May 2004 from http://citeseer.ist.psu.edu/kalfoglou02initiating.html
13. Laudon, K and Laudon J. (2002): Managing Information Systems, Managing the Digital Firm. 7th edition. Prentice-Hall International.
14. Malhotra, Y. and Galletta, D. (1999): Extending the Technology Acceptance Model to Account for Social Influence: Theoretical Bases and Empirical Validation. Proceedings of the 32nd Hawaii International Conference on System Sciences.
15. Malhotra, Y. (2004): Integrating Knowledge Management Technologies n Organizational Business Processes: Setting Real Time Enterprises to Deliver Real Business Performance. Journal of Knowledge Management. Special Issue on "Knowledge Management and Technology" Q3.
16. Park, J. and Wentling, T.: Cost Analysis of e-learning: A Case Study of a University Program (2002). Retrieved on May 2004 from http://learning.ncsa.uiuc.edu/papers/AHRD2002_wentling-park.pdf.
17. Peter, L.: Through the Looking Glass: Student Perceptions of Online Learning. The Technology Source. September/October (2001).
18. Preece, J.; Rogers, Y. and Sharp, H. (2002): Interaction Design: Beyond Human-Computer Interaction. New York, NY: John Wiley & Sons.
19. Rosson, M.B. and Carroll, J.M. (2002): Usability Engineering. Scenario-based Development of Human-Computer Interaction. Morgan Kaufmann Publishers.
20. Organ, D. and Bateman, T. (1991): Organizational Behavior. Fourth edition. Irwin.
21. Reeves, B. & Nass, C. (1996): The media equation: How people treat computers, television, and new media like real people and places, Cambridge University Press.
22. Robinson, E.: Maximizing the Return on Investment for Distance Education Offerings. Online Journal of Distance Learning Administration. Fall (2001), Volume IV, Number III.
23. Rentroia-Bonito, M.A. and Jorge, J. (2004): Towards Predictive Models for e-Learning: What Have We Learned So Far? In C. Ghaoui (Ed.). E-Education Applications: Human Factors and Innovative Approaches. Information Science Publishing 220-234.
24. Rosenberg, M. (2001): e-Learning Strategies for Delivering Knowledge in the Digital Age. McGraw-Hill.
25. Sadik, A.: Directions for Future Research in On-line Distance Education. Turkish Online Journal of Distance Education. October (2003), Volume 4, Number.4
26. Winograd, T. and Flores F. (1986): Understanding Computers and Cognition. A New Foundation for Design. Ablex Publishing Corporation. USA.
27. Xia, W. and Lee, G. (2004): Grasping the complexity of IS development projects. Communication of the ACM, N° 5 69-74.

2. P-Dinamet: A Web-Based Adaptive Learning System to Assist Learners and Teachers

Miguel Montero[1] and Elena Gaudioso[2]

[1] ICT Consultant in the Centro de Profesores y Recursos de Albacete, Physics and chemistry. Lecturer in Spanish Compulsory Secondary Education. Avenida de España 12, 02002 , SPAIN E-mail: mmontero@jccm.es

[2] Artificial Intelligence Department, E.T.S.I Informática UNED. C/Juan del Rosal, 16, 28040 Madrid SPAIN E-mail: elena@dia.uned.es

In this chapter we present P-Dinamet, a web-based educational system that recommends to the learner certain actions within a Physics course. We review an earlier version of the system and the pedagogical model behind it, then present an extended system that adapts to both learners and teachers. The suggestions are made according the information stored in learners models containing information about personal data, preferences and knowledge level. These models are presented to the teacher along with information about the learners' performance in the course and the suggestions to be presented to the learners. The teacher can modify the learners' models and provide feedback to the adaptation mechanism.

2.1 Introduction

The New Technologies are quickly increasing their presence in our society, leading a process of change in the relation forms that settle down in their inside. The number of users who use Internet and the electronic mail in their daily life is doubtlessly increasing. Thus, the number of users who, for example, consult their banking account through the Web, look for information on any subject of their interest, send electronic mails or comprise of virtual communities is constantly increasing. All these changes that little by little break through in our society, do not do likewise with the same facility in our classrooms. The daily educational practice has varied little from how it was done with our parents or we ourselves. Teachers often teach the way they were taught, and this practice can be the worse enemy for a change in education. Thus, the role of the teacher continues being the one of a trasmisor of knowledge and therefore the one of a "*infallible expert*" in the field of his/her knowledge, whereas the student plays a role of mere receiver of the knowledge, and therefore a totally passive one in the learning process. These expositions, from our point of view, make difficult the accomplishment of authentic significant learning, and what is worse, they move away from a fundamental objective: preparing our students for life, or in other words, they don't facilitate the student "*learning to learn*". These drawbacks are mainly due to the lack of adaptation of the technologies used in the classroom. Usually these

M. Montero and E. Gaudioso: *P-Dinamet: A Web-Based Adaptive Learning System to Assist Learners and Teachers*, StudFuzz **178**, 23–48 (2005)
www.springerlink.com

systems are rigid and they do not take into account the needs and preferences of both teachers and learners.

To overcome these limitations, nowadays, web-based adaptive learning systems are a reality. The main goal of adaptation in educational systems is to guide the students through the course material in order to improve the effectiveness of the learning process. The development of these systems is not a completely new research area. It inherits techniques and goals from both Intelligent Tutoring Systems [30] and Adaptive Hypermedia Systems [6].

In these systems, several adaptation tasks have been implemented for very different domains (adaptive navigation support, detection and diagnosis of students' mistakes,...). At first, Intelligent Tutoring systems implemented adaptation tasks by means of several rules that were predefined by an expert on the course domain and according certain heuristics. This approach is similar to the one applied in expert systems and has been redefined now to include more dynamic adaptation mechanisms. The techniques used are mainly from the machine learning research area [3].

Nevertheless, the adaptation tasks to be performed and the techniques used to achieve them depend mainly on the domain. The goals and requirements of a web-based adaptive learning system for secondary education differ from those of a system for postgraduate courses. In the first case more guidance is needed while the students of postgraduate courses prefer more freedom to study materials.

In addition, little attention has been paid on the assistance to teachers in these environments. This is partially due to the fact that educational systems are intended to be used outside classrooms. Nevertheless, and as it has been mentioned before, these systems must be included in school and from this point of view, it is essential to help the teachers in this task. A teacher who uses a web-based educational system needs to know how the students perform in the course and what kind of difficulties have been found. In order to fix the potential problems in the course, it is advisable also that the system recommend to the teacher how to improve the course, e.g. if the system detects that a particular exercise is better than other to help a particular student. These recommendations must be presented to the teachers so they can accept, reject or modify them. These modifications should be included in the system so it can take into account them for the next recommendation. Note that in order to provide adaptation, the system must monitor the students interactions to construct the knowledge bases in which the recommendations will be based on. These knowledge bases are also called models for adaptation and include information about students (students' models), about pedagogical issues (pedagogical models) and contents and resources (content and resources models).

With these goals in mind, our objective is to construct a Web-based adaptive Learning system directed to the teaching of Dynamics within the area of the Physics in secondary education. Our system is called P-Dinamet (Personalized Dinamet) and it is based on a system called Dinamet. Dinamet is a hypermedia system implemented according to an appropriate instructional design based on our educational experience [23]. Its elements, structure and design responds to a constructivist conception of learning, setting up scenes where the student develops an active and interactive learning. The underlying educational model in Dinamet is based on both pedagogical and design criteria which can be applied to other types of experimental matters (e.g. Mathematics).

With P-Dinamet our goal is to provide an adaptive response to both teachers and students. Thus, in P-Dinamet are embedded explicit rules for instructional guidance to students that can be modified by teachers, and mechanisms to provide more dynamic recommendations based on students interactions with the system, such us, intelligent analysis of students' interactions and recommendation of instructional strategies.

In addition, and to allow the collaboration between teachers we allow the inclusion of new educational resources, so they can be also considered for recommendation in P-Dinamet.

To illustrate our approach we first introduce certain pedagogical aspects that should be considered in the development of virtual educational systems for secondary education. Next, we will describe our previous experiences with Dinamet, its pedagogical and contents design, its learning elements and the conclusions obtained and that justify the development of the adaptive version of the system, P-Dinamet. We then describe the architecture of P-Dinamet, the models considered for adaptation and the adaptation tasks considered at the moment. We conclude with a description of related work and our conclusions and directions for future work.

2.2 Pedagogical Models in Virtual Educational Systems for Secondary Education

It has been already highlighted that the educational domain in which the system will be used determines heavily the design and the use of system.

Usually, students in virtual educational environments interact with the materials without any restriction. Nevertheless, in secondary education, students might be more closely guided. In this domain, virtual environments are often used as a complementary tool to traditional classroom, thus the goal of these systems is distinct from those used in distance learning.

In this sense the pedagogical models behind virtual educational environments will influence the interaction of the students with the system. This model can be implemented in two ways. On the one hand it can be the tutor who controls the interactions of the students and on the other hand can be the interface of the system which controls them (such as, for example, by means of roles and rights). The more open the system is, more freedom have the students to interact with the system. This fact can be a disadvantage in secondary education since the students can feel lost in the environment. Some systems, try to overcome these limitations with certain support to students [10]. Other systems try to overcome these limitations structuring the system so that the students are implicitly guided by the system [7].

In secondary education both manners become relevant. Students must be encouraged with a system that allow them to interact with very different types of materials. In addition, the system should guide the students in their study. As a complementary tool to traditional education, the system should provide the teacher with certain support to help her to control how the students are using the system, which problems are they facing, which kind of errors are making, etc...

The pedagogical models behind virtual educational environments can be specified in very different ways. A very extended approach is the use of instructional design languages based on standards like for example IMS-Learning Design [20]. These languages allow the specification of learning materials of very diverse nature (e.g. multimedia) by means of certain characteristics (e.g. difficulty level). In this paper we show that the learning materials provided by P-Dinamet are characterized by means of several attributes [21].

The elements, structure and design of Dinamet responds to a constructivist conception of learning, setting up scenes where the student develops an active and interactive learning. Its underlying educational model is based on both pedagogical and design criteria which can be applied to other types of experimental matters.

2.3 Previous Experiences with Dinamet

2.3.1 Pedagogical Design

The elements, structure and design applied in Dinamet respond to a constructivist conception of the learning, and tries to establish surroundings where the student can make an active and interactive learning. The paper of the teacher (the system itself) will be to facilitate the necessary aid so that the learner constructs its own learning processes. Our model will come near to Roger Schank's natural model of learning [27] which tries to favor the active and significant learning. This model is based on the principle of once a question about a subject in which we are interested has been made, then we will be prepared to learn the answer. The underlying educational model as well as the pedagogical and designing criteria that has been developed in Dinamet are based on basic pedagogical criteria that can be applied to another type of experimental subjects.

2.3.2 Contents Design in Dinamet

The contents must be significant and functional for the students [1]. The text that appears in screen will have to be the significant unit. We try to make it not too long, considering that reading in the screen is more tiresome than reading in printed paper. The extensions, explanations or relations on concepts will appear in additional links. In agreement with the exposed ideas, the pages in which theoretical concepts are developed will be constructed according the following guidelines:

- They will not be mere developments of the contents but they will try to generate questions from the student, starting off from the previous knowledge or preconceptions that the student can have on the subject at issue.
- The contents will help to the student to know the learning objectives that they can reach at every precise moment.
- We will propose in-between small activities or questions that help to the reflection, generates questions on the subject that is being developed, as well as serves to center the attention on the basic aspects developed.
- Separated from the main body, a section of observations that will complete the exhibition will appear. It will prompt the student to widen the content they are studying or will relate to other aspects that can be of their interest.
- The activity that the student unfolds in the construction of the knowledge cannot be solved by a single student. On the contrary it will need the aid of other mates (collaborative learning). This is the reason why diverse devices that facilitate this contact between the pupils will be provided. Among them we emphasize: notes and remarks on the material that composes Dinamet, forums, activities designed with shared common objectives...

To conclude this section it is convenient to remark that the design followed in Dinamet is a flexible structure that allows the student's navigate through the contents according to their own interests, motivations or learning rhythms. A more complete overview of the different elements of Dinamet will be described in the following sections.

2.3.3 Learning Activities in Dinamet

Considering the constructivist framework and the natural learning model mentioned before, the general directions that we follow in the design of the learning activities in Dinamet are:

- They appear like an accessible goal for the student, considering their actual competence level of and promoting their advance with the necessary aid [29]. In our surroundings this external aid can be provided by means of the communication with other companions or the tutor by telematic means (news, forums, chats, ...).
- They produce a cognoscitive conflict in the learner, facilitating the mental activity so that they can establish relations between the new contents and the previous knowledge including in the students' cognitive structure.

- They ask from the student the accomplishment of predictions and hypotheses, allowing their later verification.
- They are motivating and foment a favorable attitude in relation to the learning of new concepts
- They show a certain degree of realism that facilitates the implication of the student in their resolution as well as their possible application in the real life.
- They must help the student to acquire skills directed to *"learning to learn"*, to be autonomous in learning.

2.3.4 Learning Elements of Dinamet

Dinamet is something more than a collection of HTML pages. It is a proposal of education through the Web, dealing with what has been called Interactive Learning Environments. The resources that we include are:

- ***Educational portal:*** This is the entrance page to Dinamet and from it we may gain access to the different resources which are offered to the student. It is a page of obligatory reference for the students.
- ***Theory pages***. In them, the student can find theoretical information, in addition to commentaries, explanations and links to resources related to the basic subject that the learner is studying. In Dinamet, each theory page is divided in three different areas (see **Figure 1**):
    - o *Complementary resources:* which link to exercises, tests, recommendations, additional links and any other resource related to the theory shown in the page. It is included also an annotation tool to allow learners to send some comments to the page that can be viewed by other students.
    - o *Main content:* In the main body of the page it is shown the theorical content that the student should study.
    - o *Observations and reflection:* In this area, we include several observations or additional information that can help learners to reflect on the contents of the page.

- ***Workshops***. They are special type of exercises. Workshops have been designed with the objective that the student can create his/her own exercises, choosing the unknown factor of the problem, and introducing those values for the different variables that he considers suitable then the system offers him several possibilities and results, that can be very interesting for the student (see **Figure 2**).
- ***'Step by step' Exercises***. These exercises present to the learners a problem similar to the one being solved in the workshop but it is presented in a special form. Each step in the resolution process is presented separately (for example, in the first place it is tested if the student comprehends the formulation of the problem) . The goal of these

type of exercises is two fold. On the one hand they allow the students reflect on the process of resolution and on the other hand, they allow the system to diagnose the potential problems that a student may face in making a particular exercise.

- ***Exercises***. In these pages several types of directed exercises can be made to apply the concepts directly and mathematics formulae that the pupil has studied. In each one of them, there is a connection to the Theory Page to where the student can return to read the treated subject.

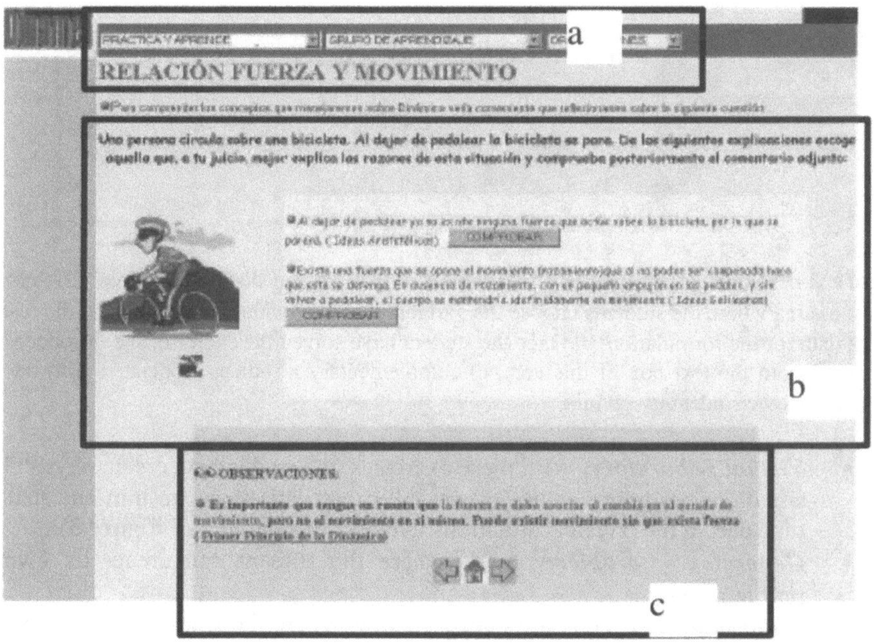

Figure 1. Example of a theory page in Dinamet where the three different areas are pointed out: a) complementary resources, b) main content and c) observations and reflection.

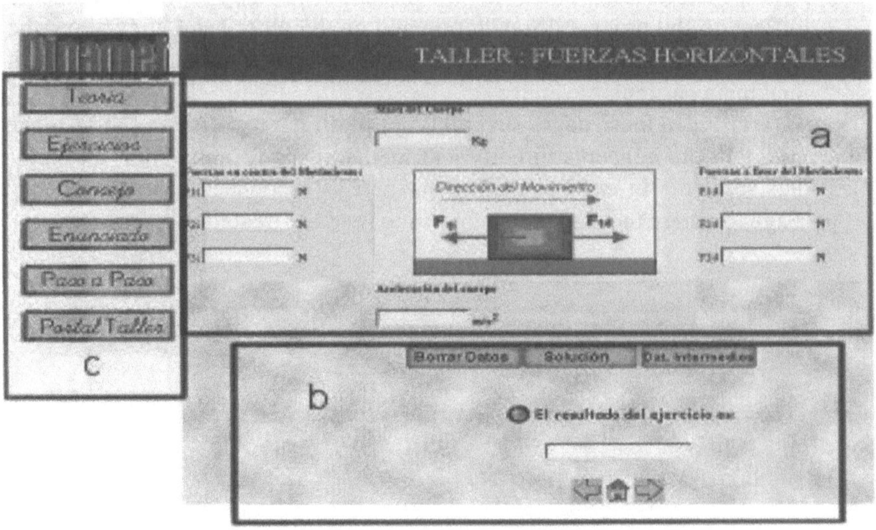

Figure 2. Example of a Workshop in Dinamet where we can distinguish three different areas: a) area where the students choose the variable that they want to calculate and the rest of the data for the formulation, b) Once the student have solved the exercise, she introduces the solution in the text box of this area, c) complementary resources associated with the variable that the student is solving.

- ***Virtual Laboratory***. With these exercises, the student can create physical situations assigning values to variables that will appear to him and that take part in the physical situations that are considered (see **Figure 3**).
- ***Generator of problems type*** where the student can create its own problems type.
- ***Autoevaluation.*** They are questionnaires where the student will be able to verify his sufficiency on these contents or on the contrary its doubts about them.

From a pedagogical point of view it is interesting to comment that the theoretical development is coherent with the exposed pedagogical principles. In order to avoid distractions but without resigning the motivation and dynamism in many of the elements of Dinamet certain animations are shown that are put into operation when the student places the mouse on them.

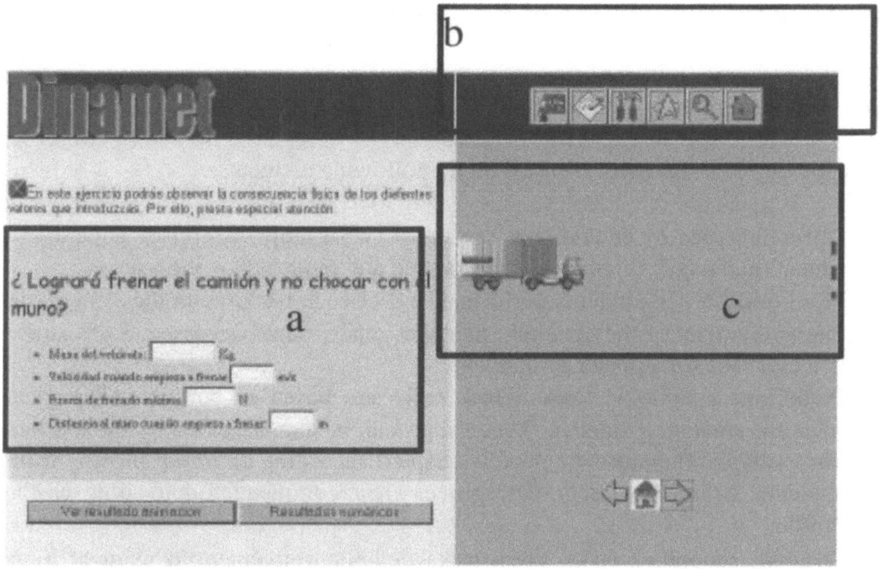

Figure 3. Example of a virtual laboratory where we can differentiate three areas: a) where it is introduced the data to be used in the simulation, b) Associated complementary resources b) Where the simulation is shown to the students.

2.3.5 Achievements and Limitations

Once Dinamet was implemented we carried out a preliminary evaluation with real users. The results obtained validated its effectiveness. It was also tested that Dinamet can be easily extended (since it is based on individual resources, it can be added any new resource or html page as desired) and that it allows an active learning since the students have access to any resource. One of the main achievement of the system is that the richness of the resources motivate students and help them to assimilate the contents.

Nevertheless, the results obtain also show how problems of erroneous preconceptions on the part of the student or poorly understood contents, take the student to retain wrong concepts in relation to the study material. It is necessary, therefore, to enable the system to give the students more sophisticated recommendations so that they face this type of situations. These recommendations will be made according the information stored in the models for adaptation. In addition the tutors did not have any information about how the students performed with the system. In the next sections we present how we plan to support the teachers in the system. This will be done presenting the learners' models and the recommendations to the tutor who will be allowed to modify both the models and the recommendations.

2.4 General Description of P-Dinamet

2.4.1 Introduction

Our approach in P-Dinamet is based on the following elements:

- Different types of educational resources (workshops, exercises generators, educational pages,...) that can be accessed and modified by the teacher.
- Monitoring of the students performance. For each resource in the system, the interaction data of the students (mistakes made, pages accessed, ...) is stored in a database for a posterior analysis.
- Predefined adaptation tasks. These tasks are based on certain pedagogical rules for students guidance. Thes e rules can be also accessed by the teacher. For example, the entrance page is adapted according an initial profile of the students, so different sets of resources are recommended depending on the profile.
- Dynamic adaptation tasks. These tasks are being implementing using artificial intelligence techniques and is in charge of help the teachers in the management of the course.

2.4.2 Architecture

P-Dinamet is composed of a Web server (Apache) connected to a relational database (MySQL) and a set of PHP scripts allowing management of the interaction with the data model. To support the interaction of the students with the different resources, we combine PHP code and JavaScript.
At the moment, the problem solving and analysis of students' solutions are embedded in pages combining PHP and Javascript. We will illustrate this within the following scenario. A student wants to practice with a workshop in P-Dinamet. Then she chooses a variable to calculate and the rest of data to compose a formulation. Once the exercis e has been solved the student fills in the solution and clicks the *'response'* button (see **Figure 2**), the JavaScript code corresponding with this exercise analyze the student response and then compares it with the correct result. All this data is sent to the database by the php code and thus all the interaction of the student with the exercise is stored in the database. We will see later in the paper, that all this information is then analyzed and incorporated in the student models. **Figure 4** illustrates this scenario.

Although the analysis made is embedded in P-Dinamet and thus, it is domain dependent, the architecture shown is exportable to other experimental matters. For this, it is needed an authoring tool allowing the creation of these pages. From our point of view, the use of JavaScript functions allow us to provide the students with an immediate feedback without overloading the server too much.

Once the exercise has been solved all the information regarding the activity of the students is stored in the database. This occurs also with any page in P-

Dinamet. In the next section, a more detailed description of the data gathering can be found.

2.4.3 Data Gathering

We have presented a system that implicitly includes certain support to students. Nevertheless to be really adaptive, P-Dinamet has to dynamically adapt its response to learners taking into account their performance in the course.
This implies that P-Dinamet have to monitor the students' interactions with the system in order to update the knowledge that the system has about the learners. This phase is common to adaptive systems and includes the gathering and preprocessing of interaction data to construct and update the models for adaptation. In P-Dinamet we capture every interaction of the students with the system. A description of the data gathered follows:

- **Students trace**: Each time a learner logs in P-Dinamet a session is initialized. For each session, a student trace includes every page view. Along with date and time of visit, we register also the type of page visited (theory, exercise,...) and the preceding page (this would allow us analyze the navigation paths of learners and detect conflictive situations).
- **Workshops**: P-Dinamet does not only register that a student have visited a particular workshop but it stores all the information related to the interaction of a student with the workshop, such as unknown variable to be calculated, data of the formulation, solution given by the student, if the student has solved the problem correctly, if the student have used some of the helps provided by the workshop,... This will allow us tracing closely the process that the student has followed in the resolution of the exercise.

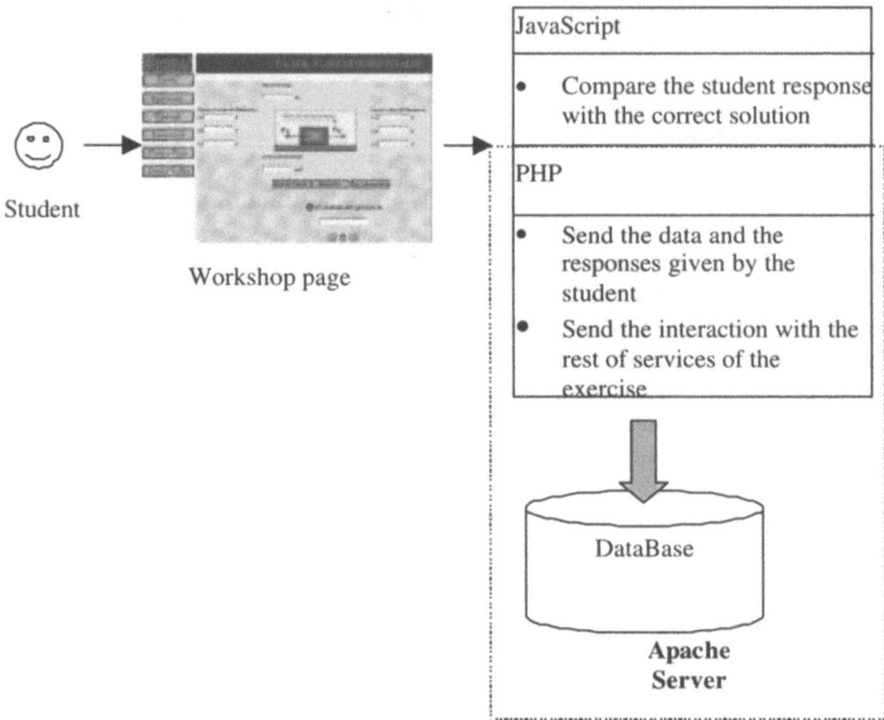

Figure 4. Scenario illustrating the architecture of P-Dinamet

- **'Step by step' exercises**: It is stored the response given by the student in each step of the problem. It will allow us to identify the potential problems that the student has in the resolution of a particular exercise.
- **Generator of problems**: Similar to the workshops, P-Dinamet stores if the solution given by the student is correct and if the student have used some of the helps provided by the generator of problems.
- **Autoevaluation**: It is stored the punctuation of the student in the whole test. In addition and in some autoevaluation tests, it is stored if the grade corresponds to conceptual issues (theory) or to practical issues (exercises).

This information is stored in the P-Dinamet database which is normalized and have 53 tables. Once the data is stored in the database, it is necessary to infer certain high level information about the students. For example, the level of knowledge of a student in a particular concept of the course, taking into account the theory pages accessed and the exercises made.
To make this step it is necessary to analyze the data stored in the database. This process is included in the so-called data preprocessing step, and allow the system to construct and update the models that will be used in adaptation tasks. These models are described in the following subsection.

2.4.4 Representation of P-Dinamet's Knowledge

Introduction

Adaptation tasks are based on the information that the system have about the students and stored in the models. Several models are usually considered in educational environments: student model, domain
model and pedagogical model.

Besides personal data, the student model contains information about the students understanding of the domain. Depending on the information considered we can distinguish three main different types of student models: *overlay* student models, *differential* student models and *perturbation* student models [19]. In overlay student models the student's knowledge is assumed to be a subset of the expert's knowledge and the goal of tutoring is to enlarge this subset. It does not cater for misconceptions or bugs that the student may have or acquire during tutoring. The differential student model is an extension of the overlay model, now knowledge is separated into that which the student has been exposed to and that which the student has not. The perturbation model caters for knowledge possessed by the student that is not present in the expert domain knowledge, in this sense, it extends the experts knowledge with the addition of a bug library (containing the most common errors that students make).

The domain model includes information about the contents that should be taught. It is composed by concepts of the course and the relationships between them (prerequisites, examples, consequence,...). It is necessary also, to include in this model the existing mapping between concepts and the materials of the course. For example, in an educational hypermedia system an html page may correspond with a single concept in the course [4].

Finally, pedagogical model include information about instructional strategies, feedback mechanisms, assessments, ... This model is sometimes embedded into domain model.
The information and representation of these models depend on the domain and goals of the educational systems. In the following subsections we will describe the models considered in P-Dinamet.

Student models in P-Dinamet

We have presented before different types of student models. The choice of the type of student model to be used is left to the system's designer. Not always a very exhaustive student model is necessary better than a simpler one [28]. In [26] it is summarized a general opinion that:

> *"detailed user models do not necessarily enhance the capability of an intelligent tutoring system ... good teaching can do without a detailed user model,*

> *because in good teaching serious misconceptions are avoided, and errors will be repaired on the spot ... it is debatable whether the cost of constructing very detailed, complex user models that are runnable and have to be maintained all the time is worthwhile in terms of the gain in teaching efficiency."*

In our case we have opted for an overlay model. This model allow us to specify the knowledge level of the students in the concepts of the course. Thus, for each concept in the course and for each student we store if the user have learnt or not the concept. In addition and as we will see in the description of the domain model, a concept can be learnt only by reading theory pages or by making exercises related to the concept. Thus, we also consider if the concept have been acquired practicing or only studying. It may happen also that an exercise makes use of a particular concept, then if a student makes this exercise correctly, then we assume that the concept is known by the student. This approach have been already used in previous successfully systems, like for example, [7].

Among the overlay model, we plan also to include in the student model some metadata that can be reused in other experimental matters. For example, if the student have difficulties in the formulation of the problem, or in the units,... this will allow to generalize some problematic situations that can be arise in other related matters. The acquisition of this metadata is not trivial, since it depends of numerous factors.

The student model in P-Dinamet also contains the following information:

- Personal data: Mainly used to identify users
- Academic data: From previous academic data we can infer the tendency to study of each student
- Computer skill level: To prevent situations in which students have problems because of the interface we store the skill of the students in the use of computers.
- Background knowledge: It comprises several attributes with information about the level of knowledge that the students have in several concepts that are used in the course. For example, in our case and for the domain of Dynamics we store the following information: Concepts and operations with vectors, basic concepts of kinematics, basic mathematical operations, previous misconceptions...
- Learning style: Learning style can be defined as the different ways a person collects, processes and organizes information. This kind of information helps more effectively adaptive systems to decide how to adapt its navigation and its presentation, thus enhancing the student learning. As usual, the systems modeling learning styles acquire this information explicitly inquiring the students through one of the existing psychometric instruments. At the moment, to model the student's learning style we have adopted the Felder-Sylverman model [16].

All this information is acquired by an initial test that it is filled in by students when registering in the system. Once acquired, P-Dinamet automatically assigns each student an initial profile. These profiles are then used to provide certain support that will be described in the following section. The profiles identified are similar to those usually considered in domain of Intelligent Tutoring Systems, namely: novice, medium and advanced. Novice students are those that present lack of background knowledge in the initial test. Medium students have sufficient background knowledge but it is the first time that study the course. Finally, advanced students have good background knowledge but need to reinforce their knowledge in the course.

The profile assigned to a student can be further updated when she improves her knowledge level. Nevertheless, this is not trivial and we are considering several options that are presented later in this paper.

Once constructed, the student model will be used to guide the students through the course materials taking into account their knowledge level and personal needs. It will be also used to select the most appropriate instructional strategy for each student.

Domain model in P-Dinamet

Domain model comprises the concepts that it is going to be presented to the students. It does not suffice to simply enumerate the concepts of the course. It is necessary to represent the relationship between concepts, such as, prerequisites or desired curriculum sequence [6]. This model it is necessary for adaptation tasks such as, analyzing students responses or recommending instructional strategies.

Acquiring the domain knowledge is a task that requires a major portion of the time and effort when building an ITS. Researchers have been exploring ways of facilitating the authoring of domain models. To this end several authoring tools have been implemented such as, TANGOW [11] or AHA [4]. Nevertheless the use of these tasks impose certain restrictions in the material to be used in the course. Nowadays it is increasing the use of educational standards that allow the specification of the course (e.g. IMS Learning Design) [20].

In our case, the domain model, although reusable in other experimental matters, presents certain particular characteristics. The concepts in P-Dinamet can be of different types:

- Prerequisite: This category includes those concepts that must be known by the students before starting the course.
- Background: This category includes those concepts that the students know from their daily life. This is specially important in experimental matters and more specifically in physics.
- Initial: This category includes basic concepts that may be studied first . These are specially important for novice students.
- Core: This category includes fundamental concepts that constitutes the course.

- Advanced: Once the core concepts are assimilated additional concepts can be presented. These concepts are intended for advanced students but can also be presented to students in medium profile.

Each concept has associated several indicators that allow the system to asses if the students have assimilated the concept. An indicator is in turn associated with an exercise or a theory page. If a student visit the theory page and/or make the exercise correctly, then the system assumes that the student knows the concept.

The relationships between the different concepts in a course that we consider are:

- Prerequisite: Concept A is prerequisite for Concept B if it is mandatory to study Concept A first before studying Concept B.
- Example: It is used to complement or clarify concepts. Usually it may happen that before studying an example it is needed to study the concept, so it is implicitly included a prerequisite relationship between the concept and the example

The elements described in the domain model are illustrated in **Figure 5**, where an example of several concepts are presented among the relationships between them.

Core concept: *Newton's Second Law*
The second law states that the acceleration of an object is dependent upon two variables - the unbalanced force acting upon the object and the mass of the object

Prerequisite Concepts: *Force* and *Acceleration*
Relationship: *Prerequisite*

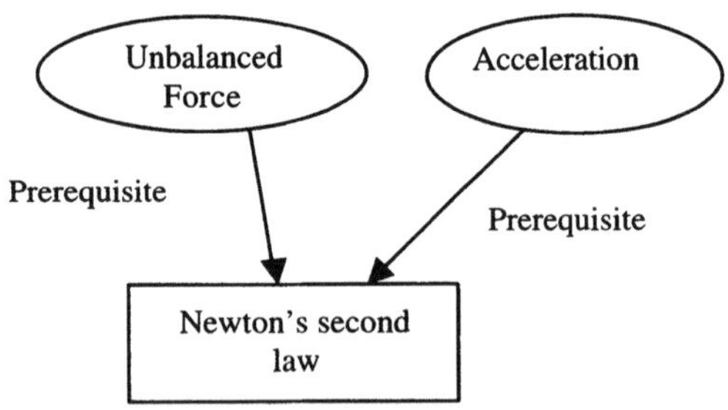

Figure 5. Example of three concepts in P-Dinamet and the relationships between them. In the left side of the figure, it is shown three concepts of the course, the type of one of the concepts is *core* and the rest of concepts are prerequisites. The relationships between them is *prerequisite.* In the right side it is graphically shown how these concepts are related.

In this domain model each concept can be associated with any of the resources that P-Dinamet offers and that have been described before (e.g., workshops, theory pages, exercises or autoevaluation tests). In addition we have associated with each resource certain characteristics that compose the so-called *meta-data for Learning Resources* in P-Dinamet. These metadata are defined according a very extended standard for Learning Object Metadata [21].

For each learning resource we specify the following main attributes: *type of resource* (workshop, theory page,...), *level of difficulty*, *learning style*, *objective* and *level of interactivity*. These features are further used to recommend a student a specific resource (for example, we recommend to novice students learning resources with a low level of difficulty). In addition, these attributes allow us to include external resources to P-Dinamet. For example, if a teacher have a collection of theory pages that can be useful to their students, she can easily incorporate them into P-Dinamet with only specifying the attributes for this new resource. In this way P-Dinamet will consider this new resource as another course material. To distinguish between external and internal learning resources, we have included also a special attribute called *character*.

Pedagogical model

It includes the information needed to guide students through the course. It contains also information about how to update the student model (for example, not update the level knowledge of a concept if the student has made a severe mistake). It allow to make decisions about the selection of an appropriate topic for the student, selection of an appropriate learning resource, and so on. During these processes the system communicates with the student and pedagogical models in order to get relevant information. At this moment, the pedagogical model in P-Dinamet includes information about:

- What type of information is needed to be presented to students according their profile.
- How can the system modify the overlay model of each student according her interaction with the materials.

At this moment this is accomplished by means of predefined rules (as shown in **Figure 6**) but as it is described in the following section we plan to make this process more dynamic by including more elaborated techniques mainly from the Artificial Intelligence research area.

```
IF
        Student.profile=Novice
THEN
        Level of difficulty: {easy, very easy}
        Type of resources: {exercises, theory pages}
```

Figure 6. Example of a rule in the pedagogical model. This rule specifies that the resources to be presented to a novice student should be exercises or theory pages with a level of difficulty easy or very easy

2.5 Adaptation in P-Dinamet

2.5.1 Introduction

We consider an adaptation task any support provided by the system to learners and teachers taking into account learners' personal characteristics and knowledge level.

To allow the system to guide the students through the course material and to assist the teacher, several adaptation tasks have been planned in P-Dinamet: Direct guidance through the course materials and support to the teacher (analysis of the students performance and recommendations about corrective actions).

Usually, adaptation tasks in adaptive systems are transparent to the users. Once an adaptive system is implemented and evaluated, end users can not modify the adaptation provided. This is a disadvantage in educational domains, since the teachers may want to modify some aspects of adaptation to her learners. In P-Dinamet, two type of adaptation tasks are considered: *static* and *dynamic*. Static adaptation tasks are those based on predefined rules (for an example, see **Figure 6**). Adaptive educational systems usually make use of these kind of rules. Nevertheless, it is hard to foresee every situation that can arise in the course. To help teachers in the management of the course, several dynamic adaptation tasks are performed. These tasks will be described in the following sections.

2.5.2 Static Adaptation Tasks in P-Dinamet

Since our domain is secondary education, it becomes necessary a direct guidance to students. This implies that P-Dinamet recommends to students the most suitable resources taking into account the information stored in the students' models. These recommendations are based on specific instructional strategies and the profile considered for each student. For example, for novice students it is advisable that the system suggests only easy resources with low interactivity level.

These adaptation tasks are based on predefined rules that can be modified for each tutor. At the moment, these rules are only based on the profile of the students.

The first time a student logs in P-Dinamet she must register and fill in an initial questionnaire. Her responses are used to infer her initial profile. As we have mentioned before, this profile are further updated as the student interacts with the system. At the moment this actualization is made according predefined rules based on the knowledge level of the student in the course, but this process is not trivial and it must be reconsidered to provide more dynamic actualization.

Once registered, the main page of P-Dinamet is adapted according to the students' profile. For example, for novice students only theory and help pages are recommended (see **Figure 7**).

Now, students can interact with the different resources. P-Dinamet will then recommend certain pages or resources according to the knowledge level of the student. These recommendations are presented in a special section called *My Recommendations.* These recommendations will be constructed on the fly whenever a student requests them. They are based on instructional rules that again are predefined and can be modified by the tutor.

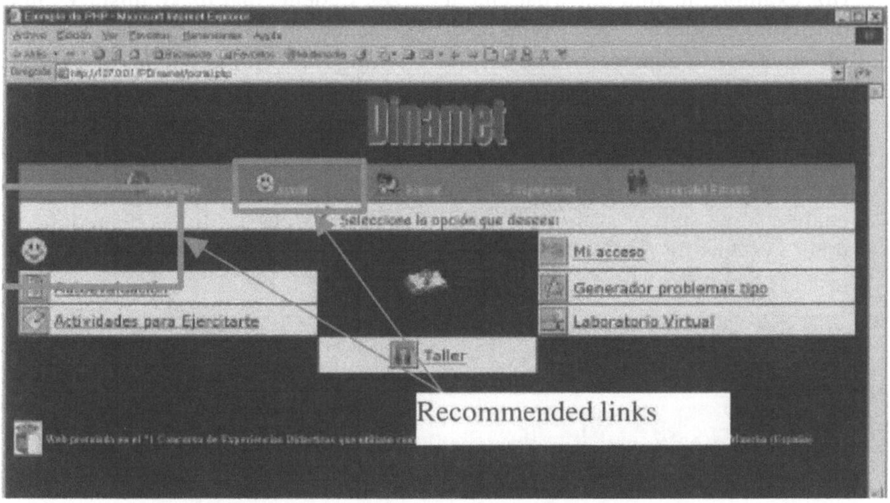

Figure 7. Entrance to P-Dinamet adapted to students with profile novice

To modify the rules described above, teachers log in P-Dinamet with special accounts and there they can access to the so-called *teachers module*. There, teachers can add new resources to the system by only assigning them the attributes specified in the last section (level of interactivity, difficulty, ...). These rules are modified by entering or modifying fields in a form. For example, in **Figure 8** can be seen how a teacher can modify the rules corresponding with the main page of P-Dinamet. Although the screenshot is in Spanish it can be seen that all the modifications are done by means of web forms.

2.5.3 Dynamic Adaptation Tasks in P-Dinamet

These tasks complement the static adaptation tasks described before. It comes up from the lack of coverage of predefined rules for every situation that could arise during the course. These tasks are mainly intended to help the teacher in the management of the course and following there is a brief description of each of them:

- *Diagnosis of student models*: When describing student models we have presented some metadata that allow the generalization of some problematic situations that can arise in the course. We plan to implement this with Bayesian networks like the ones proposed in [10].

- *Intelligent analysis of students' interactions*: In secondary education classes do not have too many students. Nevertheless the same system can be used by several groups and thus we can take advantage of all this information. We plan to use data mining techniques to analyze the interactions of the students (for example, to find out what characterizes students that make mistakes in a particular exercise). This would help the teachers to recommend new learning resources or to characterize better the existing ones (for example, to find out that an exercise is not as difficult as expected).
- *Recommend instructional strategies*: By analyzing student models and interactions the system may recommend certain navigation paths, certain exercises, ... These strategies should be presented to the teacher who can change some step or assign priorities to some exercises... It has been done in other systems by means of intelligent planning techniques [14], classification tasks [12] or case based reasoning [17].

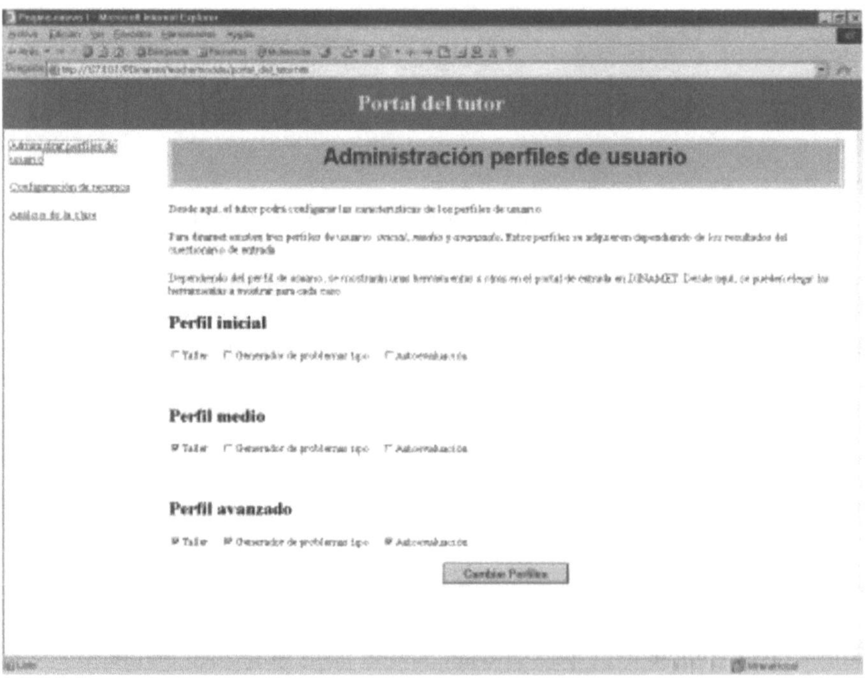

Figure 8. Teachers module in P-Dinamet.

2.6 Related Work

From the decade of the 50's, the use of the computers in the education has been considered an important field of activity. Computer Aided Instruction (CAI) was born in the 90's inheriting directly methods of work of the education programmed proposed and developed by psychologist B.F. Skinner at the end of years 50. This theory defended that the people worked on the basis of stimuli, affirming that for a same stimulus always the same answer would be obtained. Traditional programs CAI can be considered like the electronic descendants of conventional text books, since like them, they are organized statically of such form that contain so much the knowledge domain as the tutorial knowledge of the teachers. A book cannot respond to unexpected questions of the reader, and although a book can have different levels of reading, differentiation between basic materials for one first reading or of extension for later readings, a book cannot modify on the fly the displayed knowledge.

These educational systems were clearly inadequate in those areas where it is necessary that the student put into play his creative or critics attitudes. It is by that, around 70's and in parallel to the development of the computer based educational systems, it is began to develop first Intelligent Tutoring Systems (ITS), which used formalisms and techniques from the Artificial Intelligence research area to assure a better presentation of the knowledge that it was wanted to distribute. Their objective was to eliminate the little instructional flexibility of the traditional CAI systems where the presentation of the materials, the sequence in the learning, the correction of the exercises, was predefined.

We have already seen that this type of systems is based on: knowledge of the domain (expert model or domain model), a knowledge of the student (student model), and knowledge of the instructional strategies that will be applied (instructional or pedagogical model). Normally these modules operate of interactive form and they communicate through a central module that usually denominates surroundings module (interface) [30].

Although many of the results obtained in the area of Intelligent Tutorial Systems continue still effective, these systems have not been as successful as it was expected at first. The main reason, could be that these systems are applied to very particular domains. With the boom of Internet, these systems were really difficult to extend mainly due to the great amount of students who could interact with the system. In this situation the development of pedagogical models was really a hard work, because it implies the definition beforehand of every situation that could arise during the course.

These disadvantages, along with the evolution of new educational approaches when arising Internet as a tool for education, have caused that it was evolved of the first ITSs to the so-called web-based ITS [5]. Some of the most well-known systems are: CALAT [24] , ELM-ART [7], PAT Online [25] , Albatros [22].

With the new types of material that the web offers the area of development of web-based ITS have inherited techniques from the adaptive hypermedia research area.

Adaptive hypermedia systems (AHS) [6] are hypermedia systems that adapt its contents and links to the necessities and individual characteristics of the users. These adaptation tasks are usually performed by means of Artificial Intelligence techniques. In education, traditional hypermedia systems offer the same didactic answer to different students whose objectives and knowledge are different, which at least, puts in doubt the pedagogical value of the suggested materials. This situation is palliated in the adaptive hypermedia educational systems that construct a model of objectives, preferences and knowledge of each individual user and use these models during the interaction of the student with the system, adapting its response to the students' needs [15].

According to [6] three techniques of adaptation can be identified:

- Adaptive presentation support: it is an adaptation at content level. The information that the student study, can vary in details, explanations,...
- Adaptive Navigation support: the objective is to adapt the navigation of the user. This adaptation is made by hiding, reorganization some links, ...
- Adaptive Collaboration Support: the objective is to support the users in their collaboration with the other users, for example, finding peers to collaborate within the forum.

Finally, it is important to mention that due to the boom of Internet, 1996 can be considered as a crucial date in the AHS research area. At this moment the development of web-based AHS means the popularization of these types of technologies. Two very well know adaptive hypermedia systems are Interbook [8] and AHA [4].

In spite of the proven success of this type of systems, we have already commented that the domain of application and the level of education influences in the implementation decisions. In this sense, most of the implemented systems, are focused on university education.

The students of secondary education, in general, do not have procedure adapted to the study of the experimental matters like Physics, also they lack, of background knowledge that allows them to acquire new concepts, and create an adapted conceptual structure. It makes necessary in this level a greater degree of support than in higher levels.

We found adaptive systems for secondary education solely in the field of the intelligent tutorial systems, like for examp le, the system ANDES [18]. It is an intelligent tutoring system for Physic. ANDES is intended primarily to be used by students to do their physic homeworks. All other instructional activities, such as lectures, recitations and laboratories, are performed in the classroom. In ANDES, the most important components are: *The Homework Assignment Editor* (it is used by instructors to create homework assignments) and *The Tutor*. The Tutor is used by students to do their homework. It consists of the following modules: *The workbench* (where the student selects activities and makes them), *The Helper* (which tries to find out the plan or goals the student is pursuing as the student makes an activity. It offers help upon request but sometimes it may offer unsolicited help).

ANDES is directed fundamentally to the learning of knowledge of procedural type. From our didactic conception, we think that it is interesting to relate this type of contents with those of conceptual type. On the other hand, we think it necessary that the professor can obtain data of the work made by student to be able to make those decisions that he considers adequate for the improvement of the learning process. In this sense, it seems to us interesting that the system, at the request of the professor, can incorporate different resources, to be used when they are necessary.

At the moment, the rest of systems for secondary in Physics can be framed fundamentally in three types: simulation tools of physical processes, CAI and non adaptive hypermedia systems. The simulations play an important role to understand models that can be potentially difficult for the students. It is important to point out that for inexperienced users in the handling of the simulations or with little knowledge on the subject that is being taught, these types of exercises can not have all its pedagogical validity since these students may not be able to select the values of the most significant parameters. This is way it is needed to consider establishing a didactic support in the form of recommendations to the students.

In P-Dinamet, we combine different types of materials along with recommendations to the students to support their study (using the above mentioned adaptive technologies). In addition it is provided support to the teachers so that instructional strategies are recommended and it is offered the possibility of adapting the materials of P-Dinamet. Finally, there is the possibility of including external educational resources so that they can be also managed by the system.

To conclude, it is important to point out that few evaluation experiments have been performed with adaptive educational systems [13]. Our final goal with P-Dinamet is the validation of the pedagogical model implemented in the system in actual educational situations comparing its validity with traditional educational systems, as well as, to validate it as a tool to support the teaching staff.

2.7 Summary and Future Work

The inclusion of the new technologies in the classroom is not trivial. They must help both the students and teachers to promote active learning. To this end, systems that are often used in the classroom support students and guide them through the course materials.

The degree of support provided depends on the domain in which the system is applied. In this paper we have presented the special requirements that the domain of secondary education imposes. We have also described the significance of providing support to teachers to help them in the use of educational systems in the classroom.

Once exposed our goals, we have presented Dinamet, a web-based educational system directed to the teaching of Dynamics within the area of the Physics in secondary education. We have seen that its pedagogical model and design of learning activities are adequate for our domain. Nevertheless, the results obtained from preliminary evaluations shown that it is necessary to enable the system to give the students and teachers more sophisticated recommendations so that they face problematic situations that arise in the course. To this end we have developed an adaptive version of Dinamet called P-Dinamet.

P-Dinamet guide the students through the course materials. The recommendations are based on students' profile that are acquired from the responses given in the initial test. These profiles are then further updated depending on the knowledge level of the students. These recommendations are implemented by means of predefined rules that can be seen and modified by teachers through web forms.

Due to the lack of coverage of predefined rules for every situation that could arise during the course we have presented more dynamic adaptation tasks, such us, for example, intelligent analysis of students' interactions.

Although the design of P-Dinamet can be used also in other experimental matters (such us mathematics), the present version is implemented only for the course of Dynamics. To make P-Dinamet less domain dependant, three main directions lie ahead for future work:

- Dynamic actualization of students' profiles. For example, using machine learning techniques.
- Development and evaluation of dynamic adaptation tasks. We plan to start implementing the recommendation of instructional designs.
- Development of an authoring tool that allow to implement new materials for the course.

The two first points are related to the so-called *open learners models* [9]. It is an emergent topic within the research area of Intelligent Tutoring systems. It is based on the idea of showing to the students the models that the system has constructed. In this way students can both reflect on their knowledge level and help the system to construct accurate student models (by letting the students modify some of the values of the student model). Some approaches also let the teachers to modify the student models or combine the values provided by the system with the teachers point of view [2].

Acknowledgements

We would like to express our deep gratitude to professor Jesús G. Boticario for his support in this project. We would also like to thank the rest of lecturers of schools of secondary education belonging to the "*Centro de Profesores y Recursos de Albacete*" for their collaboration, support and ideas. Finally, we would thank to the "*Consejería de Educación de la Junta de Comunidades de Castilla-La Mancha*" for the award given to Dinamet

References

1. Adell, J. (1995). La navegación hipertextual en el World-Wide Web: implicaciones para el diseño de materiales educativos. II Congreso de Nuevas Tecnologías de la Información y Comunicación para la Educación. Universitat de les Illes Balears, Palma, Spain. 22, 23 y 24 November.
2. Beck, J., Stern, M., and Woolf, B. P. (1997), Cooperative Student Models, In *Artificial Intelligence in Education* B du Boulay & R Mizoguchi (eds), IOS Press, Amsterdam, 127-134.
3. Beck, J., and Woolf, B. P. (2000) High-level student modeling with machine learning. In *Proceedings of the Fifth International Conference on Intelligent Tutoring Systems*, pages: 584-593.
4. Bra, P. and Calvi, L. (1998). AHA! an open adaptive hypermedia architecture. *The New Review of Hypermedia and Multimedia*, 1(4):115–139.
5. Brusilovsky, P. (1995). Intelligent tutoring systems for world-wide web. In *Proceedings of Third International WWW Conference (Posters)*, R.Holzapfel, editor, pages 42–45, Dramstadt, Fraunhoofer Innstitute for Computer Graphics.
6. Brusilovsky, P (1996). Methods and Techniques of Adaptive Hypermedia. *User Modeling and User-Adapted Interaction*, 6: 87-129. Kluwer Academic Publishers.
7. Brusilovsky, P., Schwarz, E. and Weber, G. (1996). ELM -ART: An intelligent tutoring system on world wide web. In *Proceedings of the Third International Conference on Intelligent Tutoring Systems*, pages 261–269, Montreal, Springer Verlag.
8. Brusilovsky, P. , Schwarz, E. and Weber, G. (1996). A tool for developing adaptive electronic textbooks on www. In *Proceedings of WebNet96*, pages 64–69, San Francisco, CA, October 1996. World Conference of the Web Society.
9. Bull, S. & McKay, M. (2004). An Open Learner Model for Children and Teachers: Inspecting Knowledge Level of Individuals and Peers. To appear in *Proceedings of Intelligent Tutoring Systems 2004*, Springer-Verlag, Berlin Heidelberg.
10. Bunt A. and Conati C. (2003). Probabilistic Student Modelling to Improve Exploratory Behaviour. *Journal of User Modeling and User-Adapted Interaction*, vol 13 (3), pages 269-309
11. Carro, R.M., Pulido, E. and Rodríguez, P. (2000). TANGOW: A system for adaptive distance learning through Internet. In *Computers and Education in the 21st Century*, M. Ortega y J. Bravo (eds.), pp. 127–136. Kluwer Academic Publisher.
12. Castillo, G., Gama, J. and Breda, A.M. (2003). Adaptive Bayes for a Student Modelling Prediction Task based on Learning Styles. In *Proceedings of the 9th International Conference on User Modelling.* P. Brusilovsky, Corbett, A. and F. de Rosis (eds.). Lecture notes in Artificial Intelligence num. 2702, Springer Verlag. pp 328-332.
13. Chin, D. (2001) Empirical evaluations of user models and user-adapted systems. *User Modelling and User Adapted Interaction*, 11(1):181–194.
14. Darío, N., Méndez, D., Jiménez, C. and Guzmán, J. A. (2004). IA Planning for Automatic Generation of Customized Virtual Courses. To appear in *Proceedings of the ECAI094 Workshop on Planning and Scheduling: Bridging Theory to Practice.* Valencia (Spain). 22-23 August.

15. Eklund, J. and Brusilovsky, P. (1998). Individualising Interaction in Web based Instructional Systems in Higher Education. *In proceedings of the AUC Academic Conference 98*. University de Melbourne, Melbourne. Australia. September 27-30, pp. 27-30
16. Felder R. M., Silverman L. K. (1998). Learning and Teaching Styles. *Engineering Education* 78(7), 674–681. Preface: Felder R.˙ M., June 2002. Available in: http://www.ncsu.edu/felder-public/Papers/LS-1988.pdf
17. Funk, P. and Conlan, O. (2002). Case-based Reasoning to Improve Adaptability of Intelligent Tutoring Systems. In *Proceedings of the Workshop on Case-based Reasoning for Education and Training*. pages 15-23. September 4th. Robert Gordan University, Aberdeen. Scotland.
18. Gertner, A.S., Conati, C, and VanLehn, K. (1998). Procedural Help in Andes: Generating Hints Using a Bayesian Network Student Model. In *Proceedings of the 15th National Conference on Artificial Intelligence*, pages 106-111. Madison, Wisconsin.
19. Holt, P., Dubs, S. , Jones, M. and Greer, J.E.(1994). The state of student modelling. In *Student modelling: The key to individualized knowledge-based instruction* J.E. Greer and G.I. McCalla (eds.). NATOASI Series (Vol. 125), pp. 3–38. Springer-Verlag.
20. IMS LD (2003). *Learning Design Specification* version 1.1. 2003. http://www.imsglobal.org/
21. IMS LOM (2001): *IMS Learning Resource Metadata specification* version 1.1.2. http://www.imsglobal.org
22. Lai, M. C. , Chen, B. H. and Yuan, S. M. (1995) Toward a new educational environment. In *Proceedings of 4th International WWW Conference*, pages 221-230. Boston, USA, December.
23. Montero M., Gaudioso E (2003). DINAMET: A constructivist Approach to Adaptive Educational Hypermedia Systems. In *Proceedings of the Second International Conference on Multimedia and Information & Communication Technologies in Education (m-ICTE 2003)*. Badajoz, Spain.
24. Nakabayashi, K. (1996). An intelligent tutoring system on the www supporting interactive simulation environments with a multimedia viewer control mechanism. In *Proceedings of WebNet96*, pages 366-371, San Francisco, CA, October . World Conference of the Web Society
25. Ritter, S. (1997). PAT Online: A model-tracing tutor on the world-wide web. In *Proceedings of Workshop Intelligent Educational Systems on the World-Wide Web at AI-ED97 P. Brusilovsky, K. Nakabayashi, and S. Ritter, editors, 8th World Conference on Artificial Intelligence in Education*, pages 11–17, Kobe, Japan.
26. Sandberg, J.A.C. (1987). Coaching Strategies for Help Sy stems: *EUROHELP. The Third International Conference on Artificial Intelligence and Education*. AICOM, vol. 0, pp. 51--53
27. Schank, R.C. y Cleary, C. (1995). *Engines for education*. Lawrence Erlbaum Associates, Hillsdale, New Jersey.
28. Self, J. (1988). Bypassing the intractable problem of student modelling. In *Proceedings of Intelligent Tutoring Systems '88*, Gauthier, G., and Frasson, C., eds., pages 18-24.
29. Vigostky, L. S. (1979). *El desarrollo de los procesos psicológicos superiores*. Barcelona: Crítica, D.L.
30. Wenger, E.(1987). *Artificial Intelligence and Tutoring Systems*. Morgan Kaufmann.

3. Intelligent Agents that Learn to Deliver Online Materials to Students Better: Agent Design, Simulation and Assumptions

Leen-Kiat Soh, Todd Blank, and Lee Dee Miler

University of Nebraska, Computer Science and Engineering, 115 Ferguson Hall, Lincoln, NE 68588-0115 USA, email: {lksoh, tblank, lmille}@cse.unl.edu

In this chapter, we discuss an integrated framework of case-based learning (CBL) in an agent that intelligently delivers learning materials to students. The agent customizes its delivery strategy for each student based on the student's background profile and his or her interactions with the graphic user interface (GUI) in our system, and based on the usage history of the learning materials. The agent's decision-making process is powered by case-based reasoning (CBR). To improve its reasoning process, our agent *learns* the differences between good cases (cases with a good solution for its problem space) and bad cases (cases with a bad solution for its problem space). It also *meta-learns* adaptation heuristics and the significance of the cases' input features. We have also built a simulation to comprehensively test the learning behavior of our agent. Our design of agent learning, adaptation of a solution through CBR, and simulation is based on a set of domain-specific and independent assumptions.

3.1 Introduction

Traditionally, learning materials are delivered in a textual format and on paper. For example, a learning module on a topic may include a description (or a tutorial) of the topic, a few examples illustrating the topic, and one or more exercise problems to gauge how well the students have achieved the expected understanding of the topic. The delivery mechanism of these learning materials has traditionally been via textbooks and/or instructions provided by a teacher. A teacher, for example, may provide a few pages of notes about a topic, explain the topic for a few minutes, discuss a couple of examples, and then give some exercise problems as homework. During the delivery, students ask questions and the teacher attempts to answer the questions accordingly. Thus, the delivery is interactive: the teacher learns how well the students have mastered the topic, and the students clarify their understanding of the topic. In a traditional classroom of a relatively small size, the above scenario is feasible. However, when e-learning approaches such as distance learning and asynchronous learning are involved, or in the case of a large class size, the traditional delivery mechanism is often not feasible.

In this chapter, we propose an intelligent agent that delivers learning materials based on the usage history of the learning materials, the student static background

L.-K. Soh, T. Blank, and L.D. Miler: *Intelligent Agents that Learn to Deliver Online Materials to Students Better: Agent Design, Simulation and Assumptions* , StudFuzz **178**, 49–80 (2005)
www.springerlink.com

profile (such as GPA, majors, interests, and courses taken), and the student dynamic activity profile (based on their interactions with the agent). The agent uses the profiles to decide, through case-based reasoning (CBR) [10], which learning modules (examples and problems) to present to the students. Our CBR treats the input situation as a problem, and the solution is basically the specification of an appropriate example or problem. Our agent also uses the usage history of each learning material to adjust the appropriateness of the examples and problems in a particular situation. We have built an end-to-end intelligent agent infrastructure, with an interactive GUI front-end that is active, an agent powered by CBR and capable of learning, and a mySQL multi-database backend. Note that due to its adaptivity and modularity, CBR has been extended to several prominent Artificial Intelligence areas such as multiagent systems [14] and data mining [15]. Further CBR has been popular in interactive applications such as recommender systems [16], conversational systems [17], and software validation and testing [18]. In some designs, only the case-based retrieval mechanism of CBR is utilized, while the adaptation and case learning are not. Usually, such a design involves retrieving partially matched instances for a query. In our design, we utilize the full functionality of CBR.

Our intelligent agent is specifically designed to deliver learning modules. Each learning module consists of three components: (1) a tutorial (e.g., "What is permutation?"), (2) a set of related examples, and (3) a set of exe rcise problems to assess the student's understanding of the topic. Based on how a student progresses through the learning material and based on his or her profile, an agent will choose the appropriate examples and exercise problems for the student. In this manner, the agent will customize the learning material. Most software tutors or learning delivery mechanisms are able to customize the learning material for different students, with or without agent-based technology. The uniqueness of our design stems from (1) its modularization of the course content, the delivery mechanism, and the knowledge bases, (2) its utilization of true agent intelligence where an agent is able to learn how to deliver its learning module better, and (3) its own accountability of usefulness for evaluation. The agent uses a suite of knowledge bases: instructional strategies—e.g., which problems to select—encoded as cases, adaptation heuristics, and similarity weights for CBR, the quality of learning materials, how the learning materials are used as a set of attributes, and the models of students based on their on their aptitude and motivation background and dynamic interaction with the agent.

The overall goal of our project is three-fold. First, we want to develop courseware in introductory computer science classes where the class size is usually large and class diversity is high. For this situation, our goal is an agent that is able to customize the delivery of learning materials for each student in a flexible way. Such an agent would be of significant benefit. Second, we want to establish a flexible, easy-to-use database of courseware and intelligent agent system, including operational items such as student profiles, delivery success rates, etc., and educational items such as learner model, domain expertise, and course content. This will allow teachers and educators to monitor and track student progress, the quality of the learning materials, and the appropriateness of the materials for different student groups. Specifically, we plan to build an agent capable of adapting its delivery of examples and exercise problems to students' real-time behavior and historical profile, learning useful delivery strategies, and performing self-monitoring and evaluation tasks. With the ability to self-monitor and evaluate, the agent can identify how best to deliver a learning module to a particular student type with distinctive behaviors.

In this Chapter, we focus on the following. First, we describe the design of our agent in terms of the overall architecture from the graphical user interface (GUI) to the learner model that we adopt. Also, we will describe the various modules used in the case-based reasoning (CBR) approach that lends our agent its intelligence and machine learning capabilities. We will also focus on a rather comprehensive simulator built to simulate students interacting with our agent. This simulation supports our testing of the system. Second, we describe in details three sets of assumptions used in our design. These assumptions underlie the design of our agent learning, the simulation, and the case-based adaptation mechanism. These assumptions integrate the educational domain applications and expertise into the agent design. Finally, we provide some performance analysis of the system. These results lay the groundwork for the eventual deployment of the intelligent agent system. Indeed, the system is being deployed in real classrooms in the Fall semester of 2004.

In the following, we first discuss some related work in the area of intelligent tutoring systems. Then, we present our Intelligent Learning Materials Delivery Agent (ILMDA) project: its goals and framework. Subsequently, we propose the CBR methodology and design. We then present the set of assumptions for our design, in terms of agent learning, simulation, and adaptation in Section 3.5. In Section 3.6, we describe the implementation of the ILMDA software and a comprehensive simulation. Next we report on our preliminary experiments. Finally, we conclude.

3.2 Related Work

Research strongly supports the user of technology as a catalyst for improving the learning environment [12]. Educational technology has been shown to stimulate more interactive teaching, effective grouping of students, and cooperative learning. The cost effectiveness of using technology in terms of time saving has been estimated by several studies as about 30%. At first, professors can be expected to struggle with the change brought about by technology. However, they will adopt, adapt, and eventually learn to use technology effortlessly and creatively [8].

Kulik and Kulik [11], Bangert-Drowns et al. [1], and Baxter [2] defined several types of computer-aided education systems. In Computer-Assisted Instruction, the system provides drill and practice exercises and tutorial instruction. In Computer-Managed Instruction, the system evaluates and stores student performance and guides students to appropriate instructional resources. In Computer-Enriched Instruction, the system satisfies student requests such as solving a mathematical equation, generating data, and executing programs.

As summarized in Graesser et al. [7], intelligent tutoring systems (ITSs) are clearly one of the successful enterprises in AI. There is a long list of ITSs that have been tested on humans and have proven to facilitate learning. These ITSs use a variety of computational modules that are familiar to those of us in AI: production systems, Bayesian networks, schema templates, theorem proving, and explanatory reasoning. Graesser et al. [7] also pointed out the weaknesses of the current state of tutoring systems: First, it is possible for students to guess and find an answer and such shallow learning will not be detected by the system. Second, ITSs do not involve students in conversations so students might not learn the domain's language. Third, with ITSs, students tend to focus on the details instead of the overall picture of a solution.

There have been successful ITSs such as PACT [9], ANDES [6], AutoTutor [7], and SAM [5], but without machine learning capabilities. These systems do not generally adapt to new circumstance, do not self-evaluate and self-configure their own strategies, and do not monitor the usage history of the learning materials being delivered or presented to the students. In our research, we aim to build intelligent tutoring agents that are able to learn how to deliver appropriate learning materials to different types of students and to monitor and evaluate how the students receive the learning materials.

3.3 Project Framework

In the Intelligent Learning Materials Delivery Agent (ILMDA) project we aim to design an intelligent agent to deliver learning materials. Each learning material consists of three components: (1) a *tutorial*, (2) a set of related *examples*, and (3) a set of *exercise problems* to assess the student's understanding of the topic. Based on how a student progresses through the learning material and based on his or her profile, an ILMDA will choose the appropriate examples and exercise problems for the student. In this manner, the ILMDA will customize the learning material. Most software tutors or learning delivery mechanisms are able to customize the learning material for different students, with or without agent-based technology. Our design is modular for both the course content and delivery mechanism, utilizes true agent intelligence where an agent is able to learn how to deliver its learning materials better, and self-evaluates its own learning materials.

The underlying assumptions behind the design of our agent are the following. First, a student's behavior in viewing an online tutorial, and how he or she interacts with the tutorial, the examples, and the exercise problems, is a good indicator of how well the student understands the topic in question, and this behavior is observable and quantifiable. Second, different students exhibit different behaviors for different topics such that it is possible to show a student's understanding of a topic, say, T1, with an example E1, and at the same time to show the same student's lack of understanding of the same topic T1 with another E2, and this differentiation is known and can be implemented.

Further, we want to develop an integrated, flexible, easy-to-use database of courseware and ILMDA system, including operational items such as student profiles, ILMDA success rates, etc., and educational items such as learner model, domain expertise, and course content. This will allow teachers and educators to monitor and track student progress, the quality of the learning materials, and the appropriateness of the materials for different student groups.

Finally, we plan to build an agent capable of adapting its delivery of examples and exercise problems to students' real-time behavior and historical profile, learning useful delivery strategies, and performing self-monitoring and evaluation tasks. With the ability to self-monitor and evaluate, the agent can identify how best to deliver a learning module to a particular student type with distinctive behaviors. We see this as valuable knowledge to instructional designers and educational researchers as ILMDA not only is a testbed for testing hypotheses, but it is also an active decision maker that can expose knowledge or patterns that are previously unknown to researchers.

3.4 Methodology

Our ILMDA system is based on a three-tier methodology, as shown in Figure 1. The agent consists of a graphical user interface (GUI) front-end application, a database backend, and the reasoning module in between. A student user accesses the learning material through the GUI. The agent captures the student's interactions with the GUI and provides the reasoning module with a parametric profile of the student and environment. The reasoning module performs case-based reasoning to obtain a search query (a vector of search keys) to retrieve and adapt the most appropriate example or problem from the database. The agent then delivers the example or problem in real-time back to the user through the interface. In our agent, we use case-based reasoning (CBR) to select and adapt which examples and problems the student is given.

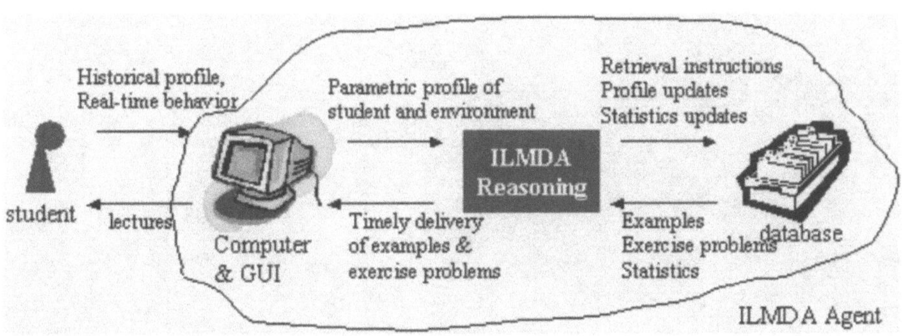

Figure 1. Overall methodology of the ILMDA system.

3.4.1 Overall Flow of Operations

When a student starts the ILMDA application, he or she is first asked to login. This associates the user with his or her profile information. The information is stored in two separate tables. All of their generally static information, such as name, major, interests, etc., is stored in one table, while the user's dynamic information (i.e., how much time, on average, they spend in each session; how many times they click the mouse in each session, etc.) is stored in another table. After a student is logged in, he or she selects a topic and then views the tutorial on that topic. Following the tutorial the agent looks at the student's static profile, as well as the dynamic actions the student took in the tutorial, and searches the database for a similar case. The agent then adapts the output of that similar case depending on how the cases differ, and uses the adapted output to search for a suitable example to give to the student. After the student is done looking at the examples, the same process is used to select an appropriate problem. Again, the agent takes into account how the student behaved during the example, as well as his or her background profile. After the student completes an example or an exercise problem, they may elect to be given another. If they do so, the agent assumes that the last example or exercise problem it gave the student did not satisfy the student's need, and tries a different solution. Figure 2 shows the interaction steps between our ILMDA agent interacts and a student. Note that a database may not contain examples/problems of all possible combinations of attributes. As a result, when the agent prescribes for an example or a problem with particular attributes, it may not find a matching example or problem. Faced with this possibility, our agent actually performs a relaxed retrieval based on fuzzy sets.

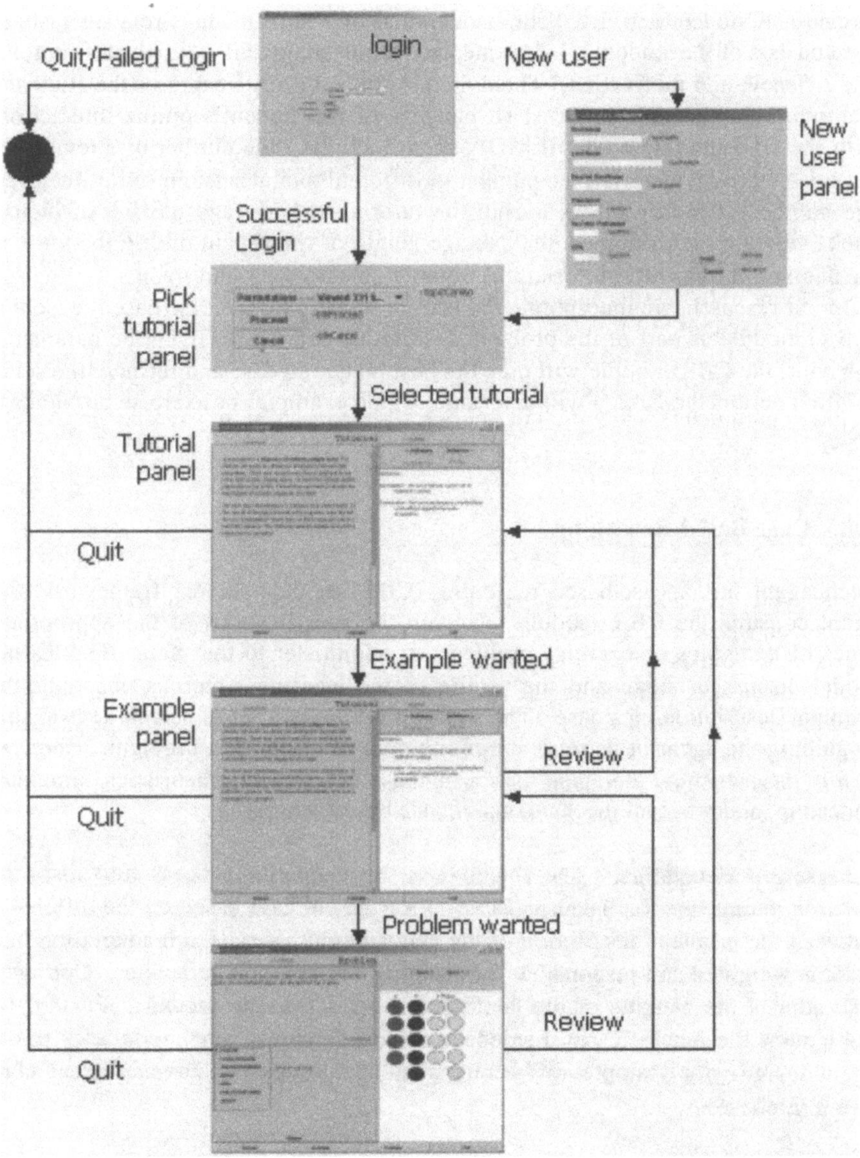

Figure 2. GUI and interactions between ILMDA and students.

3.4.2 Learner Model

A learner model is one that tells us the metacognitive level of a student by looking at the student's behavior as he or she interacts with the learning materials. We achieve this by profiling a learner/student along two dimensions: student background and student activity. The background of a student stays relatively static and consists of the student's last name, first name, major, GPA, goals, affiliations, self-efficacy, and motivation. The dynamic student profile captures the student's real-time behavior and patterns. It consists of the student's online interactions with the GUI module of the ILMDA agent including the number of attempts on the same learning material, the number of different modules taken so far, the average number of mouse clicks during the tutorial, the average number of mouse clicks viewing the examples, the average length of time spent during the tutorial, the number of quits after tutorial, the number of successes, and so on.

In our research, we incorporate the learner model into the case-based reasoning (CBR) module as part of the problem description of a case: Given the parametric behavior, the CBR module will pick the best matching case and retrieve the set of solution parameters which will determine which examples or exercise problems to pick.

3.4.3 Case-Based Reasoning

Each agent has a case-based reasoning (CBR) module. In our framework, the agent consults the CBR module to obtain the specifications of the appropriate types of examples or exercise problems to administer to the users. The learner model discussed above and the profile of the learning materials constitute the problem description of a case. The solution is simply a set of search keys (Table 1) guiding the agent in its retrieval of either an example or a problem. Here, we briefly describe how the agent uses a rulebase of adaptation heuristics, simulated annealing, and retrieval mechanisms of case-based reasoning.

Adaptation Heuristics. The objective of the adaptation process is to adapt the solution parameters (as listed in Table 1) for the old case based on the difference between the problem description of the new and old cases. Each adaptation heuristic is weighted and responsible for adapting one solution parameter. Our agent also adjusts the weights of the heuristics using a learning module. Finally, we implement the heuristics in a rulebase to add flexibility and modularity to our agent design—for example, different student age groups may have different adaptation rulebases.

Table 1. The solution parameters of a case in the ILMDA system. These parameters form the search query to retrieve an example or a problem from the database.

Solution Parameters	**Description**
TimesViewed	The number of times the material has been viewed
DiffLevel	The difficulty level of the material between 0 and 10
MinUseTime	The shortest time, in milliseconds, a single student has viewed the material
MaxUseTime	The longest time, in milliseconds, a single student has viewed the material
AveUseTime	The average time, in milliseconds, the material has been viewed by students
Bloom	Bloom's taxonomy number: knowledge, comprehension, application, analysis, evaluation, synthesis
Scaffold	Degree of scaffolding: highlights, hints, references, elaborations
Content	The stored list of interests of this material
Length	The number of characters in the course content (tutorial) of this material
AveClick	The average number of clicks the interface has recorded for this material

Simulated Annealing. The situation may arise where the adaptation process is given an old case with undesirable solution parameters—a set that has consistently led to unsuccessful outcomes. To avoid repeating the same mistake, we use simulated annealing to obtain different and possibly satisfactory solutions. Because this simulated annealing process is costly compared to the normal adaptation process, an agent does this only when it encounters a case that is retrieved often and that has a low success rate.

Fuzzy Retrieval. As mentioned earlier in Section 3.4.1, our agent uses a relaxed retrieval strategy based on fuzzy sets to obtain a partial match, to guarantee that there is always a problem or an example for the student. This strategy first attempts to find a perfect match between the prescription (i.e., the search key) and the appropriate example/problem in the database. When a perfect match is not possible, the agent relaxes the least important attribute-value pair in the search key. The relaxation is based on the membership of the value in the attribute-value pair in the fuzzy sets of attribute. For example, if the value is a member of set 1, which covers values in the range R, then the agent will search for a match in the database using the modified attribute-range pair with the rest of the search key. The agent continues with this relaxation until a match is found. As long as there is at least one example or a problem in the database, the agent is guaranteed to find one that partially matches. Thus, it is possible that the search key was prescribed correctly but the example/problem failed due to the poorly stocked database. With our design that separates the prescription from the retrieval process, an agent is able to monitor and even pinpoint the causes of possible failures of its interactions with students.

Case-Based Learning (CBL). To improve its reasoning process, our agent learns the differences between good cases (cases with a good solution for its problem space) and bad cases (cases with a bad solution for its problem space). It also meta-learns adaptation heuristics, and the significance of input features of the cases.

Our agent uses various weights when selecting a similar case, a similar example, or a similar problem. By adjusting these weights we can improve our results, and hence, learn from our experiences. In order to improve our agent's independence, we want to have the agent adjust the weights without human intervention. To do this, the agent uses simple methods to adjust the weights called learning modules. Adjusting the weights in this manner gives us a layered learning system [13] because the changes that one module makes propagate through to the other modules. For instance, the changes we make to the similarity heuristics will alter which cases the other modules perceive as similar. Our learning design is also akin to instance based learning [4] where each case is considered an instance.

In our agent design, we have four CBL modules. First, there is a case learning module that learns about good and bad cases, through both traditional methods and by using simulated annealing. Secondly, we have a similarity learning module that learns about which case attributes are important for a case to be similar as well as dissimilar. Lastly, we have an adaptation learning module that learns about the importance or quality of the adaptation heuristics. The case-learning module uses traditional CBL and a small amount of reinforcement learning. The other two modules meta-learn how to better retrieve from the database, evaluate the similarity between two cases, and adapt the solution of the best case to the current problem. This layered learning allows our agent to learn how to better serve the students.

We will discuss the design and implementation of CBL later in Section 3.6.

3.4.4 User Interface

The ILMDA front-end GUI application is written in Java, using the Sun Java Swing library. A student progresses through the tutorials, examples, and problems in the following manner, as shown in Figure 2 shown in Section 3.4.1. The student logs onto the system. If he or she is a new student, then a new profile is created. The student then selects a topic to study. The agent administers the tutorial associated with the topic to the student accordingly. The student studies the tutorial, occasionally clicking and scrolling. Then when the student is ready to view an example, he or she clicks to proceed. Sensing this click, the agent immediately captures all the mouse activity and time spent in the tutorial and updates the student's activity profile. The student goes through a similar process and may choose to quit the system, indicating a failure of our example, or going back to the tutorial page for further clarification. If the student decides that he or she is ready for the problem, the agent will retrieve the most appropriate one based on the student's updated dynamic profile.

3.4.5 Simulation

We have also built a comprehensive simulator to provide us with virtual students to use ILMDA. This simulator allows us to test the correctness and feasibility of the ILDMA, as well as to evaluate our learner and instructional models in the future. When we run our experiments, this simulator bypasses the GUI component and directly feeds the agent with simulated student background and activity profiles.

Our simulator consists of two distinct modules: a *Student Generator* and an *Outcome Generator*. The first module generates nine different basic types of students based on their aptitudes and speeds, further qualified by their motivation and self-efficacy values (Table 2) in using online learning materials. This also guides the simulation of actual student activity. The second module simulates interactions and outcomes. The assumptions used to guide the design and development of the simulator are discussed in Section 3.5.2.

Student Generator. The Student Generator module creates the *virtual* students that ILMDA interacts with. This module generates (1) all student background values such as names, GPAs, and interests, and (2) the activity profile such as the average time spent on a session and average number of mouse clicks. The underlying approach we use is to choose a random value from a Gaussian distribution qualified by the student type. As an example for aptitude, it has three Gaussian distributions. A random value is used to indicate where the student's aptitude is, given a particular aptitude type (high, average, or low). Then, given that aptitude value, its probability is determined based on the Gaussian distribution for the type. Similarly, another random value is generated and processed for speed. These two values are then averaged using different weights to generate a single unit-less value that we call the *independent parameter value* (IPV). To obtain a value for a particular problem descriptor such as the average number of mouse clicks (*aveClicks*), the IPV is mapped proportionally to the bounds of *aveClicks*. The resultant mapping yields the simulated value for *aveClicks* for one *virtual* student.

Table 2. Table of student types used by ILMDA Simulator. E.g., type 2 student has high aptitude and medium speed, high self-efficacy, and high motivation.

Student Type	Motivation	High			Medium			Low		
Self-efficacy	Aptitude/ Speed	Fast	Med.	Slow	Fast	Med.	Slow	Fast	Med.	Slow
High	High	1	2	3	10	11	12	19	20	21
	Average	4	5	6	13	14	15	22	23	24
	Low	7	8	9	16	17	18	25	26	27
Medium	High	28	29	30	37	38	39	46	47	48
	Average	31	32	33	40	41	42	49	50	51
	Low	34	35	36	43	44	45	52	53	54
Low	High	55	56	57	64	65	66	73	74	75
	Average	58	59	60	67	68	69	76	77	78
	Low	61	62	63	70	71	72	79	80	81

Outcome Generator. The Outcome Generator simulates the *interaction* and *quit* behavior for each virtual student. The interaction behavior is simulated using the results of simulated values from the Student Generator and statistics on the learning material stored in the database. To illustrate, the simulated time spent on an example (*ExampleTime*), the simulated number of clicks on an example (*ExampleClicks)*, and the simulated number of times that the student goes from the example back to the tutorial (*ExmpToTutorial)* are computed as follows:

$$ExampleTime = ((XL/AXL)*(XD/AXD))/2*SAET*RANDOM$$
$$ExampleClicks = (((XL/AXL)*(XD/AXD))/2*SAEC*RANDOM$$
$$ExmpToTutorial = (((XL/AXL)*(XD/AXD))/2*SAETT*RANDOM$$

where *XL* = the example's length; *AXL* = the average example length for this topic; *XD* = the example's difficulty level; *AXD* = the average example difficulty for this topic; *SAET* = the student's average example time; *SAEC* = the student's average number of example clicks; *SAETT* = the student's average example to tutorial clicks; *RANDOM* = a random number between .8 and 1.2.

For problems the functions that outcome generate uses are very similar to those used for the examples:

$$ProblemTime = ((XL/AXL) * (XD/AXD))/2 * SAPT * RANDOM$$
$$ProblemClicks = ((XL/AXL)*(XD/AXD))/2 * SAPC * RANDOM$$
$$ProbToTutorial = ((XL/AXL)*(XD/AXD))/2 * SAPTT * RANDOM$$
$$ProbToExample = ((XL/AXL)*(XD/AXD))/2 * SAPTE * RANDOM$$

where *PL* = the problem's length; *APL* = the average problem length for this topic; *PD* = the problem's difficulty level; *APD* = the average problem difficulty for this topic; *SAPT* = the student's average problem time; *SAPC* = the student's average number of problem clicks; *SAPTT* = the student's average problem to tutorial clicks; *SAPTE* = the student's average problem to example clicks; and *RANDOM* = a random number between .8 and 1.2.

We use the above formulae to generate the actual interaction behavior. With this and the student background, the simulator now has a complete problem description for the case-based reasoning to take place.

A student may choose to quit while reading the tutorial, or going through the exercise, or working on the problem, depending on the student's aptitude and the difficulty level of the example and problem he or she encounters. To simulate this, we have the following assumptions.

We start with a base probability that a given student type will quit any tutorial, example, or problem. We assume that high-aptitude students are less likely to quit than other students. Table 3 shows the base probabilities. The quitting rate for tutorials is assumed to be 5% lower than the base quitting rates for an example or a problem, as we assume that students are less likely to quit a tutorial.

Table 3. Base probabilities for quitting for each aptitude type. Numbers in parentheses indicate the values for tutorial.

High	**Average**	**Low**
10% (5%)	25% (20%)	40% (35%)

To decide whether a student would quit a tutorial, we simply randomly (using a uniform distribution) make a student quit according to the above base probabilities.

To decide whether a student would quit an example, we modify the base probabilities by adding to it two values: a difficulty-based modification and a time-based modification. Our rationale is that a student is more likely to quit an example when the example is more difficult than appropriate for the student's aptitude, and when the example takes longer to go through than appropriate for the student's speed. The difficulty-based modification and the time-based modification are computed as follows:

$$\textit{Difficulty Modifier} = (DL)/10 - (1 - FP)*(.7777777)$$
$$\textit{Time Spent Modifier} = .3*((XT - AUXT)/(XT + AUXT))$$

Where *DL* = difficulty level of the example; *FP* = the fixed percentage chance to quit; *XT* = student's example time; and *AUXT* = average use time for example.

To decide whether a student would quit a problem, we use the same function as above with problem time instead of example time.

$$\textit{Difficulty Modifier} = (DL)/10 - (1 - FP)*(.7777777)$$
$$\textit{Time Spent Modifier} = .3*((PT - AUPT)/(PT + AUPT))$$

where DL = difficulty level of the example; FP = the fixed percentage chance to quit; High – 10%; Average – 25%; Low – 40%; PT = student's problem time; and AUPT = average use time for problem.

To decide whether a student would answer a problem correctly, it is based on the product of the final, modified probability and a factor we call the *motivation factor* (MF). We see that a student who has made it through the tutorial, the example, and the problem is more likely to answer the problem correctly.

3.4.6 Machine Learning Modules

As described in Section 3.4.3, our agent is equipped with machine learning capabilities. It is able to track the performance of its actions, link the previous decisions that lead to those actions, and evaluate whether and how such actions should be performed in the future when it encounters similar events. There are two key learning modules: similarity and adaptation heuristics. The similarity learning module learns about how to better determine the similarity between two problems (i.e., roughly two cases). Knowing how to better determine the similarity between two problems allows the agent to retrieve more accurately old cases and solutions that better fit the new problems. The adaptation heuristics learning module learns about how to better adapt old solutions to the new problem. Knowing this allows the agent to better apply what it knows from its past experience to new situations, making it more adaptive.

Similarity Learning Module. The weights that the agent uses for deciding the similarity of two cases define what it believes are important attributes in the cases. If the weight of attribute "A" is high, and the weight of attribute "B" is low, then the agent will look for a case that has a similar attribute "A". By adjusting these weights, the agent adapts to different circumstances. The module adapts these values by systematically going through each case, and finding a separate case that has similar values (using the old weights). It then compares the outcome of each case. If the outcome is the same (i.e. both cases resulted in success, or both in failure) then it is assumed that the cases are indeed similar, and the current weights should be strengthened. If the cases have different results, then the module assumes that these cases should not be considered similar, and weakens the current weights. By continually doing this, we eventually end up with weights that accurately provide us with similar cases. Table 4 shows the attributes that factor into the similarity value.

Table 4 Input features (or attributes) of a case.

Input Variables	Description
AveExmpClicks	The average number of times the student clicks the mouse in the examples he or she has seen. (When comparing two cases to pick a problem, the number of example clicks in that session is used instead).
AveExmpTime	The average time spent (in milliseconds) per example. (When comparing two cases to pick a problem, the amount of time spent during the example clicks in that session is used instead).
AveExmpToTtrl	The average number of times the student goes back to the tutorial from the example.
AveGrade	The student's average grade on the problems.
AveProbClicks	The average number of times the student clicks the mouse in the problems he or she has seen.
AveProbTime	The average time spent (in milliseconds) per problem.
AveProbToExmp	The average number of times the student goes back to the example from the problem.
AveProbToTtrl	The average number of times the student goes back to the Tutorial from the problem.
AveSesTime	The student's average total time spent in the interface during a session.
ExmpQuits	The number of times the student has quit at the example stage.
MaxSesTime	The length of the student's longest session, in milliseconds.
MinSesTime	The length of the student's shortest session, in milliseconds.
NumExmp	The number of examples the student has seen.
NumProb	The number of problems the student has seen.
NumSessions	The total number of sessions the student has had.
ProbQuits	The number of times the student has quit at the problem stage.
Successes	The number of successful sessions the student has had.
TtrlClicks	The number of times the user has clicked the mouse during the tutorial
TtrlQuits	The number of times the user has clicked the mouse during the tutorial.
TtrlTime	The length, in milliseconds, the student spent in the tutorial.
Self efficacy	The student's Self efficacy, measured by a pre-topic quiz
Motivation	The student's motivation, measured by a pre-topic quiz

The similarity learning module adjusts the weights that the agent uses to compute the similarity of two cases. Theoretically, cases that are most similar should have similar inputs, similar outputs, as well as identical outcomes. The flow of the similarity weights learning module is shown in Figure 3. It begins by selecting the first case in the database, and finds the most similar case to it, using the current weights. Then it compares each of the attributes in the cases, to find the most similar attributes. If the cases had the same outcome, then they are, presumably, correctly identified as similar, and any attributes that they have in common should be weighted heavier in future cases. On the other hand, if the cases had different outcomes, then perhaps they were not truly similar in the first place. So the module will find the attributes the cases have in common, and reduce those weights, so they are no longer considered to be important.

Adaptation Heuristics Learning Module. The agent adapts the solution of the most similar case to the current situation. In our design, each adaptation heuristic is represented by a vector of attribute-value pairs. Each attribute is essentially an input feature of the case. The way the agent adapts the output is dependant on several "weight" variables. These variables are displayed in Table 1.

The original weight values are again defined by education experts and may be erroneously assigned or not suitable for different groups of students. The agent may find that adapting on a particular weight variable is undesirable or that it should adapt more heavily on a particular value. The adaptation learning module notices which features in the input should affect the corresponding features in the output (i.e., the search query), and adjusts the current adaptation rules set to reflect this knowledge.

The adaptation learning module learns by observing whether a solution is successful and noting the heuristics that were used to produce that solution. If the solution is not successful (e.g., the student quits instead of finishing the example), then the heuristics that were responsible for the adaptation will be penalized, and if the outcome of the case is successful then the heuristics responsible for the adaptation will be rewarded.

The module will also identify which heuristic was responsible for the worst adaptations and the best adaptations by keeping track of how much each heuristic changed the search keys over many cases. Then the module will adjust the heuristics, which are also just weights used to adapt the search attributes, based on how often and how much they contributed to good and bad adaptations.

The adaptation heuristics control how we adapt new experiences based on old cases. In order to decide if the current heuristics are good enough, we must look at the success rate of the cases. If the cases have a generally good success rate, then it will only slightly adapt the values. If the cases have been mostly unsuccessful, then the module will adapt the values more aggressively.

The adaptation heuristic learning module uses the statistics table, which stores each search value, along with how much it was adapted by, and whether or not the adaptation produced successful results, to look for which key adaptation created the most successful outcomes. It then makes changes to the heuristics accordingly.

3.5 Assumptions

In the following, we provide the high-level assumptions used as our design principles. Overall, we have three sets of assumptions:

- The agent learning: These assumptions guide how an agent should learn to improve its own performance over time.
- The simulation: We build a simulation that imitates what a student will do facing the ILMDA software, as mentioned in Section 3.4.5. A student is charac-

terized by a set of parameters, with which the simulation probabilistically determines what the student will do given a particular type of example or problem. The determination process is guided by this set of assumptions.

- The adaptation: An agent performs solution adaptation during CBR to modify a previous solution of a similar problem such that the new solution matches more closely to the current problem. The adaptation heuristics are based on the assumptions in this set.

3.5.1 Agent Learning

Our agent's reasoning is depends upon CBR and CBL. Furthermore, our agent also learns through reinforcement about its adaptation heuristics, its actions and decisions—in hope of providing challenging enough examples and problems to the students such that (1) students do not get bored by examples or problems that are too easy, (2) students do not get discouraged by examples or problems that are too difficult, and (3) students achieve the target of the learning efficiently, using minimal numbers of examples and problems. Here are our assumptions.

Assumption A1: An agent only learns cases that are different from what it already knows.

For our learning algorithm a case must be different enough from other cases or it will not be learned, i.e., added to the casebase. In a casebase that is limited in size by the time it takes the agent to search it we do not want to store cases that are too similar in the casebase. Cases that are too similar do not add much diversity to the casebase while slowing down the agent.

Assumption A2: An agent obtains a solution for a new problem based on a previous solution to a previously encountered problem.

Each adaptation rule consists of a relative, weighted difference function between the parameters of a new problem and those of a previously encountered problem. This assumption is based on CBR.

Assumption A3: Cases with similar problem descriptions will be adapted similarly.

Assumption A4: Cases with similar problem descriptions will have similar solutions.

The above two assumptions follow from the principles used in CBR. Cases with similar problem descriptions will have similar solutions resulting from the adaptation rules used by the agent. Of course, this only applies to cases using the same adaptation rules set.

Assumption A5: An agent's casebase may have inappropriate or poor cases due to noise.

Assumption A6: An agent may use simulated annealing to derive a relatively different solution from the solution of a poor case.

Assumption A7: An agent should only use simulated annealing when the best solution for a particular problem has been observed to be consistently poor. An agent is cautious when it comes to finding a new solution for an existing case.

Bad cases are cases that have been used a certain number of times and have not resulted in a successful outcome. These cases are essentially local minima (in terms of their utility) in the casebase. They have been selected for adaptation several times due to the problems that these cases cover. However, the solutions encoded within the cases are not appropriate, leading to failures. Thus, a method other than adaptation must be used when this case is selected again. This method is *simulated annealing*, a heuristic AI method that is provably good at escaping local extrema. Note that a bad case is one on which repeated adaptation has failed to result in a successful outcome. Running the adaptation on a bad case has a poor chance of producing a successful outcome. Simulated annealing is used instead, in the hope that it can escape the local minima and find a case that is not as bad.

Assumption A8: An agent should forget about a poor solution once a new solution is found. An agent is aggressive when it comes to replacing an existing old solution with a new solution, even though the new solution has not been proved to be good.

After a case has been annealed on, the original solution parameters are replaced with the solution parameters of the case most similar to the annealed case after the annealing process terminates are used. The annealed problem description then reverts to its original setup. Thus, this basically "pulls a solution out of a bag". In a casebase whose size is limited by the total search time to cover all cases, there is no point in keeping the original case after it has been annealed on. Keeping the original case would contradict ***Assumption A4*** that indicates that cases with similar problem descriptions should have similar solution parameters.

3.5.2 Simulation

Here we list the assumptions behind the design of our simulation. Assumptions I1-I2 are implicit assumptions. These assumptions underlie other assumptions. For clarity they are listed here and referred to where relevant in the simulation assumptions.

Assumptions I1: A student will click more often in an example/problem that is long and difficult than in an example/problem that is short and easy.

The justification is that students click to access support material or go back to the tutorial/example. Therefore a student clicks when he or she does not understand the material. If the example/problem is difficult the student will click more than if it was easy. If the examp le/problem is long the student will click more than if it was short. This assumption allows the agent to infer from a student's action the length and difficulty level of an exa mple/problem.

Assumption I2: The average value of an attribute associated with a student, or an example, or a problem, is a representative value of that attribute.

The assumption is valid when the average value is computed based on a representative sample of data. This assumption allows the agent to make decisions based on the average value of an attribute, cutting down processing time and storage space that would be required to deal with the actual set of attribute values.

Assumption S1: The speed at which a student goes through a learning material is not related to the aptitude of that student.

Simulated Students are given two different characteristics *speed* and *aptitude*. There are three different categories for each characteristic. For *speed* the three different categories are fast, medium and slow. For *aptitude* the three different categories are smart, average and below average. Student Type is an integer between 1 and 9 representing the 9 different combinations of *speed* and *aptitude*. See Table 2. The justification for splitting speed and aptitude characteristics comes from real world observation. When taking a test some students finish quickly because the material is easy for them to complete and others finish quickly because it is assumed that they do not care whether they complete the material. The alternative is also true, some students spend a lot of time completing the material because they are methodical and others spend a lot of time because they do not understand the material. Thus, we see that aptitude and speed as two separate, uncorrelated parameters.

Assumption S2: The likelihood of a Simulated Student quitting the tutorial is inversely proportional to the student's aptitude.

As shown in Table 3, the fixed percentage for each aptitude is the following: high aptitude – 5% for tutorials; average aptitude – 20% for tutorials; low aptitude – 35% for tutorials. It is assumed that a student with a higher aptitude will be more likely to continue with the learning material, and that a Simulated Student with a lower aptitude will be more likely to be intimidated by the tutorial and quit.

Assumption S3: The time that a Simulated Student spends on an example is proportional to the example's length and difficulty, and also proportional to student's average time spent on other examples.

The example time for a Simulated Student can be computed using a function, as shown in Section 3.4.5. It is assumed that the longer an example is, the more time, more clicks, and more example-to-tutorials there will be. Likewise, the more difficult an example is, the more time, more clicks, and more example-to-tutorial clicks there will be. RANDOM gives us a small, natural, variation.

Assumption S4: The number of times a Simulated Student clicks in an example is proportional to the example's length and difficulty, and also proportional to the student's average number of clicks in other examples.

This assumption stems out of the implicit ***Assumption I2***. Please refer back to Section 3.4.5 for an implementation of this assumption.

Assumption S5: The number of times a Simulated Student returns to the tutorial from the currently viewed example is proportional to the example's length and difficulty, and also proportional to the student's average number of example to tutorial clicks.

The example-to-tutorial clicks for a simulated student can be computed using a function, as shown in Section 3.4.5. It is assumed that the longer an example is, the more likely that a student will go back to the tutorial to refresh his or her knowledge to help him or her understand the example better. The assumption follows from the first implicit assumption, ***I1***. Based on the implicit assumption, if the current example is longer than the average example length the number of simulated example to tutorial clicks should be increased. The same is true if current example is more difficult than the average example difficulty. The number of simulated example-to-tutorial clicks should be decreased if current example is shorter and/or less difficult than the average example.

Assumption S6: The number of times a Simulated Student spends in a problem is proportional to the problem's length and difficulty, and proportional to the student's average time spent on other problems

Assumption S7: The number of times a Simulated Student clicks in a problem is proportional to the problem's length and difficulty, and proportional to the student's average number of problem clicks.

Assumption S8: The number of times a Simulated Student returns to the tutorial from the currently viewed problem is proportional to the problem's length and difficulty, and proportional to the student's average number of problem to tutorial clicks.

The above three assumptions are similar to ***Assumptions S3-S5***, respectively, and are justified similarly. For an implementation, please refer back to Section 3.4.5.

Assumption S9: The number of times a Simulated Student returns to the example from the currently viewed problem is proportional to the problem's length and difficulty, and proportional to the student's average number of problem to example clicks.

This is similar to ***Assumptions S5*** and ***S8***. For an implementation, please refer back to Section 3.4.5.

Assumption S10: The likelihood that a Simulated Student will quit an example is based on the difference between the average time students have taken to view that example and the simulated time it takes for the Simulated Student to view that example, and the example's difficulty level with respect to the student type.

The chance a Simulated Student will quit an example can be computed using the difficulty modifier and time spent modifier described in Section 3.4.5. The chance a Simulated Student will quit at the example is based on how difficult the example is compared with the student's aptitude characteristic. This is what the *Difficulty Modifier* computes. The chance to quit is also based on how long the student would spend in the example compared with the average use time of the example. This is what the *Time Spent Modifier* computes. The results of both functions summed together to determine the likelihood. In general, if the example is difficult and the student type is of low aptitude, then it is more likely that the Simulated Student will quit the exa mple. Further, the likelihood is normalized by the average amount of time students in the past have taken to view the example.

The *Time Spent Modifier* is based on the example time (***Assumption S3***) and the average time from the example. This function will be positive if the student spent less time in the example than the average time spent in the example and negative if the opposite is true.

The *Difficulty Modifier* is based on the difficulty level of the example and the student's aptitude characteristic. This function is proportional to the difficulty level since a higher difficulty level increases the chance a student will quit. The *Difficulty Modifier* is reduced more for a student with a high aptitude than a student with a low aptitude. The justification for aptitude can be found in ***Assumption S2***.

Assumption S11: The likelihood that a Simulated Student will quit a problem is based on the difference between the average time students have taken to view that problem and the simulated time it takes for the Simulated Student to view that problem, and the problem's difficulty level with respect to the student type.

This assumption is similar to ***Assumption S10*** and is also justified similarly. Please refer back to Section 3.4.5 for an implementation.

3.5.3 Adaptation

Each agent has a case-based reasoning (CBR) module. In our framework, the agent consults the CBR module to obtain the specifications of the appropriate types of examples or problems to administer to the users. The learner model discussed above and the profile of the learning materials constitute the problem description of a case. The solution is simply a set of search keys (see Table 1) guiding the agent in its retrieval of either a problem or an exercise. Here, we briefly describe how the ILMDA agent uses a rulebase of adaptation heuristics and simulated annealing to support the adaptation and retrieval mechanisms of case-based reasoning.

The objective of the adaptation process is to adapt the solution parameters for the old case (see Table 3) based on the difference between the problem description of the new and old cases. Each adaptation heuristic is weighted and responsible for adapting one solution parameter. Our agent also adjusts the weights of the heuristics using a learning module, as discussed in Section 3.4.6. Finally, we implement the heuristics in a rulebase to add flexibility and modularity to our agent design—for example: different student age groups may have different adaptation rulebases.

Assumption AD1: If a student has had many sessions of the same tutorial, and the average numbers of times of the tutorial, example, and problem viewed by the student is high, then the agent should provide an example or problem with a high number of times it has been viewed (or timesViewed). This assumption follows from that if a student in general tends to view the same materials many times, then it is likely that materials that have been viewed many times by other students are materials appropriate for the student.

Assumption AD1.2: Students with a high motivation and/or self-efficacy are better able to handle examples or problems with fewer timesViewed.

The adaptation rule used for times viewed is a dot product of two vectors of input variables and the weights from the database. The weights are summarized as follows:

$$numSession = 10, aveTtrlTime = 2, aveExmpTime = 2, aveProbTime = 2,$$
$$motivation = \text{-}2, self\text{-}efficacy = \text{-}2$$

With this assumption, we see that students who tend to view the same materials many times will be given materials that have been viewed many times by other students. The rationale here is that the materials viewed many times by other students are likely to be materials most useful; or otherwise they would not have been delivered to the students by the agent.

The negative weights for motivation and self-efficacy are designed to provide students with these higher values with examples or problems with fewer *timesViewed*. The reason for this is that these examples or problems are newer and their reliability is suspect. In this way students with higher scholastic statistics are used to evaluate new examples or problems.

Assumption AD2: If a student has a high grade point average (GPA), a high number of successes, and a high average tutorial grade, then the agent should provide an example or problem with a high difficulty level.

Assumption AD2.1: The agent also increases the difficulty level of the example or problem to be chosen slightly if the student in average spends time in viewing the learning materials.

Assumption AD2.2: The agent also decreases the difficulty level of the example or problem to be chosen slightly if the student tends to go back and forth between different sections (tutorials, examples, and problems) of a learning material.

Assumption AD2.3: If a student has a higher motivation or self-efficacy, then the agent should present the student with an example or problem with a higher difficulty.

The adaptation rule used for difficulty level is a dot product of two vectors of input variables and the weights from the database. The weights are summarized as follows:

$$gpa = 6, successes = 8, numSessions = 2, aveProbTime = 2, aveGrade = 10, aveTtrlTime = 1, aveExmpTime = 1, aveProbTime = 1, exmpToTtrl = -1, probToTtrl = -1, probToExmp = -1, motivation = 2, self\text{-}efficacy = 2$$

These assumptions follow from the intention of the agent to challenge students with different scholastic abilities. This intention is to provide an example or problem to the student, which is challenging but not impossible. Thus a student with a poor GPA, no successes and a low average grade should receive a problem with a low difficulty level while a student with a good GPA, several successes, and a high average grade should receive a problem with a much higher difficulty level.

The same correlation is true for motivation and self-efficacy. A student with a higher motivation or self-efficacy should receive a more difficulty problem to challenge them. While a student with lower values for motivation or self-efficacy should receive a problem with a lower difficulty level.

It is also assumed that if a student is observed to have spent time in viewing the learning materials, then the agent should provide a slightly more difficult example or problem to the student. We see this as the agent's decision that the student is patient and motivated enough to want to spend time and thus deserves to have a more difficult example/problem.

Further, it is also assumed that when a student flips back and forth between an example and a tutorial, between a problem and an example, and so on, it means the student is not patient and not motivated enough to carefully go through the learning materials. As a result, the agent delivers an easier example or problem to help improve the student's interest.

Assumption AD3:* *If a student has spent a lot of time in previous tutorials, examples, and problems, then the student should be given an example or problem that requires a low minimum amount of time to view.

Assumption AD3.1:* *If a student has a high motivation, then the agent should give the student an example or problem with a higher minimum use time.

Assumption AD3.2:* *If a student has a high self-efficacy, then the student should be given an example or problem with a lower minimum use time.

The adaptation rule used for *Minimum Use Time* is a dot product of two vectors of input variables and the weights from the database. The weights are summarized as follows:

$$aveTtrlTime = 1, aveExmpTime = 1, aveProbTime = 1, motivation = 2, self\text{-}efficacy = -2$$

The justification for this assumption is that some students spend more time in the learning module than other students. This could be because they read more slowly, are more methodical, or do not understand the materials but refuse to quit. The agent wants to find an example or problem that requires a very small minimum amount of time to go through as the appropriate example or problem for the student to prevent the student from getting overly frustrated.

The motivation assumption is justified by the description of the motivation characteristic. Motivation reflects, amongst other things, how long a student is willing to spend on an example or a problem. Thus a student with a higher motivation is willing to spend more time and the agent should try to match this student with an example or a problem with a longer minimum use time.

The assumption for self-efficacy is based on the description of the self-efficacy characteristic. Self-efficacy determines how fast a student is able to learn new materials. Thus a student with a higher value for self-efficacy will need to spend less time and the agent should try to match this student with an example or a problem with a shorter minimum use time.

Assumption AD4: If a student has spent a lot of time in previous tutorials, examples, and problems, then the student should be given an example or problem that requires a high maximum amount of time to view.

Assumption AD4.1: If a student has a high motivation, then the agent should give the student an example or problem with a higher maximum use time.

Assumption AD4.2: If a student has a high self-efficacy, then the student should be given an example or problem with a lower maximum use time.

The adaptation rule used for *Maximum Use Time* is a dot product of two vectors of input variables and the weights from the database. The weights are summarized as follows:

aveTtrlTime = 1, *aveExmpTime* = 1, *aveProbTime* = 1, *motivation* = 2, *self-efficacy* =-2

The justification for this assumption is that some students spend more time in the learning module than other students. This could be because they read slower, are more methodical, or don't understand but refuse to quit. The agent wants to find an example or problem that requires a very high maximum amount of time to go through as the appropriate example of problem for the student to tailor to the student's willingness to spend time in the learning materials. Comparing ***Assumption AD4*** to ***Assumption AD3***, we see a contradiction. However, upon closer analysis, we see that ***AD4*** is aimed at preventing students from getting *frustrated*, and ***AD5*** is aimed at preventing students from getting *bored*.

The same rationale is used to justify the motivation and self-efficacy weights as that used in ***Assumption AD3***.

Assumption AD5: If a student has spent a lot of time in previous tutorials, examples, and problems, then the student should be given an example or problem that requires a high average amount of time to view.

Assumption AD5.1: If a student has a high motivation then the agent should give the student an example or problem with a higher average use time.

Assumption AD5.2: If a student has a high self-efficacy then the student should be given an example or problem with a lower average use time.

The adaptation rule used for *Average Use Time* is a dot product of two vectors of input variables and the weights from the database. The weights are summarized as follows:

aveTtrlTime = 1, *aveExmpTime* = 1, *aveProbTime* = 1, *motivation* = 2, *self-efficacy* =-2

The same rationale is used to justify the weights as that used in ***Assumption AD3***.

Assumption AD6: The Bloom's taxonomy number is important when determining if the example or problem is suitable for the student. The higher the Bloom's taxonomy number the more involved the problem is. Students with low scholastic statistics such as GPA and average score should receive a problem with a lower Bloom's taxonomy number.

Assumption AD6.1: Students who have a lot of quits at the tutorial, example, or problem stage should receive examples or problems with a lower Bloom's taxonomy number.

Assumption AD6.2: If a student has a high motivation or self-efficacy, then the agent should give the student an example or problem with a higher Bloom's taxonomy number.

The adaptation rule used for *Bloom's taxonomy number* is a dot product of two vectors of input variables and the weights from the database. The weights are summarized as follows:

gpa = 6, *successes* = 6, *ttrlQuits* = -2, *exmpQuits* = -2, *probQuits* = -2, *motivation* = 3, *self-efficacy* = 3

This assumption follows from an explanation of Bloom's taxonomy. The Bloom's taxonomy number divides all problems into six categories. The categories indicate an increasing level of student involvement with the problem but not necessarily difficulty. The six Bloom's taxonomy numbers are: (1) Knowledge/Memory, (2) Comprehension, (3) Application, (4) Analysis, (5) Synthesis, and (6) Evaluation.

A problem with knowledge/memory (1) might require the student to list, define, tell, describe, identify, show, label, collect, examine, tabulate, quote, name, who, when, where, etc. This is very different from an evaluation problem (6) requiring the student to assess, decide, rank, grade, test, measure, recommend, convince, select, judge, explain, discriminate, support, conclude, compare, summarize.

According to Bloom *et al.* [3], "questions with a higher number require much more 'brain power' and a more extensive and elaborate answer." It follows that students with higher scholastic statistics should be given problems with a higher Bloom's taxonomy number. The justification for this is the same as that found in AD2, the desire to challenge students.

The desire to challenge students also necessitates the agent giving a student with a higher self-efficacy or motivation an example or a problem with a higher

Bloom's taxonomy number. As discussed above a higher Bloom's taxonomy number implies that the problem is more involved and challenging for the student.

Assumption AD8: ***If a student spends a lot of time on average in the previous tutorials, examples, and problems, then the agent should increase the length of the example or problem to be chosen. However, if the student has quit frequently in previous tutorials, examples and problems, the agent should decrease the length of the example or problem to be chosen.***

Assumption AD8.1: ***If a student has a high motivation then the agent should give the student an example or problem with a longer length.***

Assumption AD8.2: ***If a student has a high self-efficacy then the student should be given an example or problem with a shorter length.***

The adaptation rule used for *Average Clicks* is a dot product of two vectors of input variables and the weights from the database. The weights are summarized as follows:

$$ttrlQuits = -1, exmpQuits = -1, probQuits = -1, aveTtrlTime = 3, aveExmpTime = 3,$$
$$aveProbTime = 3, motivation = 2, self\text{-}efficacy = -2$$

This assumption follows from the desire of the agent to match an appropriate example or problem to each student. Some student will prefer shorter examples/problems with less explanation while other students prefer examples/problems that are longer but provide more explanation. If a student has repeatedly quit the example/problem, then the agent will present the student with a shorter example/problem. If the student spends a lot of time in the learning module on average, then the agent will present the student with an example or problem with a longer length.

The same rationale is used to justify the motivation and self-efficacy weights as that used in ***Assumption AD3.***

3.6 Implementation and Simulation

We have implemented an end-to-end prototype ILMDA system with a front-end, applet-based Graphical User Interface (GUI) that captures all user mouse activities, a CBR-powered agent, based on the assumptions discussed in Section 3.5, and a backend database system containing a set of learning materials. We have also implemented a Simulator to test the agent by running experiments on it with virtual students. The ILMDA system is written in Java (SDK version 1.4.2). Both integrated development environments (IDEs) were run under Windows 2000/XP operating systems. The backend database is a MySQL database (Version 3.23). The system connects to the database using the JDBC drivers for Java. We have also built the comprehensive simulator to provide us with *virtual* Simulated Students to use ILMDA. When we run our experiments, this simulator bypasses the GUI component and directly feeds the agent with simulated student background and activity profiles. Figure 3 shows the three phases of our simulation-based experiments.

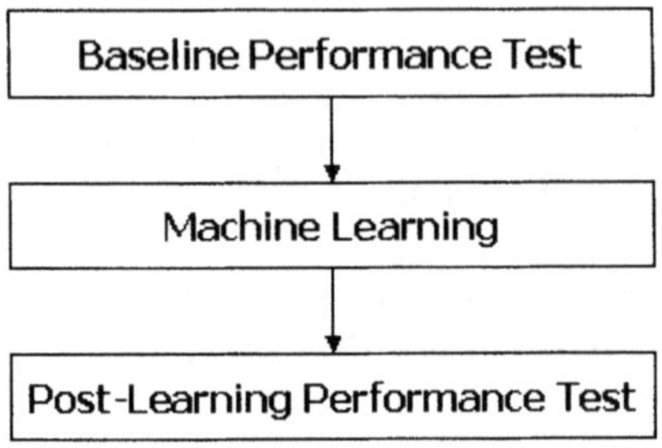

Figure 3. Three phases of the simulation-based experiments to determine the impact of agent's machine learning on the system.

The objective of our first experiments with our simulator was to determine the feasibility of ILMDA, its correctness, and the learning performance. Using the simulator discussed above, we generated 900 virtual students, with 100 students in each student type. Particularly, we focused our experiments on type-2 students, with high aptitude and medium speed. We then conducted the simulation in three steps. The first step ran the simulator for 1000 iterations with all learning modules of the ILMDA disabled. That is, no new cases were added and no weights were adjusted. The second step ran the simulator for 100 iterations with the learning modules enabled. For this experiment, we also coded the system to always learn a new case. The last step ran the simulator for 1000 iterations with the learning modules turned off. Our aim was to compare the results of the first and the third steps.

After the first and third steps, we collect the following parameters: (1) the average quitting point for students, (2) the average score of students who reach a problem, (3) the number of times each example/problem given, (4) the number of times students quit each example/problem, and (5) the number of times students answered each example/problem correctly. The average quitting point is based on the following scale: the agent receives 0 point if the student quits the tutorial, 1 point if during an example, and 2 points if during a problem. The agent receives 3 and 4 points for incorrect and correct answers, respectively.

Overall, the end-to-end behavior of our ILMDA is correct, performing as expected. Our agent is able to deliver different learning materials to different students adaptively. Our agent is also able to learn new cases and adjusts its weight. Based on the experiments, we observe that:

- The average quitting points for the first and third steps were 1.801 and 1.634, respectively. This indicates that after the agent was trained after step 2, students were more prone to quitting before reaching a problem. Indeed, only 59 students reached a problem in step 1, and 44 did in step 3. This indicates that the ILMDA failed to further the students' activity more often than before it was trained. Upon a closer look, we realize that the ILMDA was trained to provide difficult-enough examples and problems to the students. In the first step, the ILMDA identified a particular example that had a 100% success rate and learned to use another, more difficult example. Thus, from the viewpoint of the agent, we have successfully built an agent that is intelligent enough to (1) learn which problems or examples are more difficult, and (2) learn to provide those problems or examples to students who have done well in less difficult problems and examples. However, as a result, students quit more often and failed to reach the problem session. To remedy this, we are currently addressing this issue to include instructional scaffolding and hints for each example, and will report on this in the future.

- The average scores for the student who answered a problem were 0.407 and 0.568 for the first and third steps, respectively. On the surface, this indicates that the ILMDA was able to give more suitable problems to the students after the 100-iteration training received in the second step. However, cross-checking this observation with the above, we realize the following. Of those students who answered problems, more, in terms of percentage, were able to answer the problems correctly. But, more students, in terms of percentage, quit at the example or problem stage. That means the agent did indeed provide more challenging examples and problems and students who made it through to the problems tend to be able to better answer the problems correctly. This further strengthens the implication that to help students along to *attempt* to answer these more challenging examples/problems, we need to *scaffold* their interaction or understanding of these examples/problems more effectively.
- After machine learning, the ILMDA agent gave twice as many different examples to the students, better utilizing the resources it had at its disposal, comparing the results of step 1 and step 3. This shows that the agent was able to learn to retrieve different examples to give to the students, and the agent was able to learn better adaptation heuristics. An in-depth study is forthcoming to better highlight how the agent accomplishes this at different stages of its reasoning.
- After machine learning, the ILMDA gave exactly the same number of problems to the students. Coupling this with the second observation above, we see that the ILMDA was able to learn to apply more appropriate problems to different students even though it did not increase the number of problems used. Since the same set of problems were given, the students must have been given more appropriate problems, or the average scores would have stayed closer to the first value of .407. This observation gives us a wonderful insight: the ILMDA agent is able to differentiate among student types and among problem types, and strengthen its reasoning process that gives students appropriate problems, after the machine learning stage.

3.7 Conclusions and Future Work

We have described an intelligent agent that delivers learning materials adaptively to different students, factoring in the usage history of the learning materials, the student static background profile, and the student dynamic activity profile. We have built an ILMDA infrastructure, with a GUI front-end, an agent powered by case-based reasoning (CBR), and a mySQL multi-database backend. We have also built a comprehensive simulator for our experiments. Preliminary experiments demonstrate the correctness of the end-to-end behavior of the ILMDA agent, and show the feasibility of ILMDA and its learning capability. We are currently continuing with the experiments with the different student types and popularizing our database of learning materials. Future work includes incorporating complex learner and instructional models into the agent and conducting further experiments on each learning mechanism, and investigating how ILMDA adapts to a student's behavior, and how ILMDA adapts to different types of learning materials.

In the Fall 2004 semester, the ILMDA system will be deployed in the CS1 class, the first introductory core Computer Science course, with five learning materials in the following topics: File Input and Output (I/O), Event-Driven Programming, Exceptions, Inheritance and Polymorphism, and Recursion. We will incorporate these learning materials and ILMDA into the structured laboratories that are required course activities for CS1 in our Computer Science and Engineering (CSE) department. Our experiment plan includes splitting students into two groups: Control and ILMDA. The Control group will receive no ILMDA support and go through the labs regularly. The ILMDA group will use ILMDA to learn more about the above topics and then go through the labs. For each group, we will conduct pre- and post-tests. And after each lab, we will perform statistical analyses to measure the impact of ILMDA. This experiment will be underway in September 2004.

Specifically for our future work, we will continue to expand ILMDA in three directions: (1) machine learning and agent reasoning, (2) GUI support with multiple views, and (3) courseware development. We will continue to conduct fundamental research in agent cognition that involves machine learning and self-awareness. We aim to improve the agent reasoning process such that it is able to pinpoint the faulty components in its delivery of learning materials. We will add to the existing GUI package a suite of user-driven tools so that students, teachers, developers, and researchers can access student data, learning data, agent reasoning, and so on conveniently online. Finally, we will continue to develop learning materials. Our goal is to provide 14 sets of course materials for each of the four introductory CS core courses: CS1 and CS2 in terms of programming and problem solving, CS3 in discrete structures and CS4 in data structures and algorithms.

3.8 Acknowledgments

The research project is supported by the Great Plains Software Technology Initiative and the CSE Department at the University of Nebraska, Lincoln, NE. The authors would like to thank Suzette Person and Ashok Thirunavukkaras for their help during the development of the software.

References

[1] Bangert-Drowns, R., Kulik, J., and C.-L. Kulik, C.-L. 1985. Effectiveness of Computer-Based Education in Secondary Schools, Journal of Computer-Based Instruction, 12(3):59-68.

[2] Baxter, E. 1990. Comparing Conventional and Resource Based Education in Chemical Engineering: Student Perceptions of a Teaching Innovation, *Higher Education,* 19:323-340.

[3] Bloom, B. S., Mesia, B. B., and Krathwohl, D. R. 1964. *Taxonomy of Educational Objectives* (two vols: *The Affective Domain & The Cognitive Domain*). New York: David McKay.

[4] Bradshaw, G. 1987. Learning about Speech Sounds: The NEXUS Project, in *Proc. 11th Int. Workshop on Machine Learning,* Irvine, CA, 1-11

[5] Cassell, J., Annany, M., Basur, N., Bickmore, T., Chong, P., Mellis, D., Ryokai, K., Smith, J., Vilhjálmsson, H., and Yan, H. 2000. Shared Reality: Physical Collaboration with a Virtual Peer, *ACM SIGCHI Con. on Human Factors in Computer Systems*, April 1-6, The Hague, The Netherlands

[6] Gertner, A. S. and VanLehn, K. 2000. ANDES: A Coached Problem-Solving Environment for Physics, in *Proc. ITS'2000*, 133-142.

[7] Graesser, A. C., VanLehn, K., Rosé, C. P., Jordan, P. W., and Harter, D. 2001. Intelligent Tutoring Systems with Conversational Dialogue, *AI Magazine*, 22(4):39-51.

[8] Kadiyala, M. and Crynes, B. L. 1998. Where's the Proof? A Review of Literature on Effectiveness of Information Technology in Education, in *Proc. 1998 FIE Conference*, 33-37.

[9] Koedinger, K. R., Anderson, J. R., Hadley, W. H., and Mark, M. A. 1997. Intelligent Tutoring Goes to School in the Big City, *Journal of Artificial Intelligence in Education*, 8(1):30-43.

[10] Kolodner, J. 1993. *Case-Based Reasoning*. Morgan Kaufmann.

[11] Kulik, C.-L. and Kulik, J. 1991. Effectiveness of Computer-Based Instruction: an Updated Analysis, *Computers in Human Behavior* 7:75-94.

[12] Sivin-Kachala, J. and Bialo, E. 1998. *Report on the Effectiveness of Technology in Schools, 1990-1997*, Software Publishers Association

[13] Stone, P. 2000. Layered Learning in Multiagent Systems. MIT Press.

[14] Ontañón, S. and Plaza, E. 2003. Collaborative Case Retention Strategies for CBR Agents, in *Proc. 2003 International Conference on Case-Based Reasoning (ICCBR)*, 392-406, Trondhelm, Norway.

[15] Yang, Q. and Cheng, H. 2003. Case Mining from Large Databases, in *Proc. 2003* International Conference on Case-Based Reasoning (ICCBR), 691-702, Trondhelm, Norway.

[16] McGinty, L. and Smyth, B. 2003. On the Role of Diversity in Conversational Recommender Systems, in *Proc. 2003 International Conference on Case-Based Reasoning (ICCBR)*, 276-290, Trondhelm, Norway.

[17] McSherry, D. 2003. Increasing Dialogue Efficiency in Case-Based Reasoning Without Loss of Solution Quality, in *Proc. 2003 International Joint Conference on Artificial Intelligence (IJCAI)*, 121-126, Acapulco, Mexico.

[18] Gomes, P., Pereira, F. C., Carreiro, P., Paiva, P, Seco, Ferreira, J. L. and Carlos, B. 2003. Solution Verification in Software Design: A CBR Approach, in *Proc. 2003 International Conference on Case-Based Reasoning (ICCBR)*, 171-185, Trondhelm, Norway.

4. Intelligent Web-Based Computer-Supported Collaborative Learning

Vladan Devedžic

Department of Information Systems and Technologies
FON - School of Business Administration
University of Belgrade
POB 52, Jove Ilica 154, 11000 Belgrade, Serbia and Montenegro
Phone: +381-11-3950853, Fax: +381-11-461221
Email: devedzic@fon.bg.ac.yu

This chapter surveys important issues in Computer-Supported Collaborative Learning (CSCL) in the context of intelligent Web-based learning environments. It defines and explains many important issues related to intelligent Web-based CSCL and discusses them in the context of existing systems and learning environments. It only briefly touches psychological background and instructional theories that support collaborative learning. Rather than that, the chapter covers technological and practical aspects of intelligent Web-based CSCL, such as architectures of such systems and the use of modern intelligent techniques in CSCL environments (e.g., intelligent agents and Web mining). It also discusses principles used in evaluation of intelligent Web-based CSCL systems.

4.1 Introduction

Let us introduce CSCL through an example of a working Web-based CSCL environment. Figure 1 is a screenshot from COLER, an intelligent CSCL system for learning the principles of entity-relationship (ER) modeling in the domain of databases [5]. The learners using the system through the Web solve a specific ER modeling problem collaboratively. They see the *problem description* in the upper center window and build the solution in the dedicated *shared workspace* (lower center). Each learner has his/her own *private workspace* (upper right), in which he/she builds his/her own solution and can compare it to the evolving *group solution* in the shared workspace. A learner can invoke a personal coach (an intelligent pedagogical agent) to help him/her solve the problem and contribute to the group solution (upper left). In addition to such guidance, there is also a dedicated HELP button to retrieve principles of ER modeling if necessary. At any time during the problem solving, a learner can see what teammates are already connected (middle left) and can ask for floor control ("ask/take pencil" button, bottom left). When granted control, the learner contributes to the group solution in the shared workspace by, for example, inserting a new modeling element. He/she can also express feelings about other teammates' contributions through the *opinion panel* (middle

V. Devedžic: *Intelligent Web-Based Computer-Supported Collaborative Learning*, StudFuzz **178**, 81–110 (2005)
www.springerlink.com

right), and can also engage in discussion with them and with the coach through the *chat communication* window (lower right).

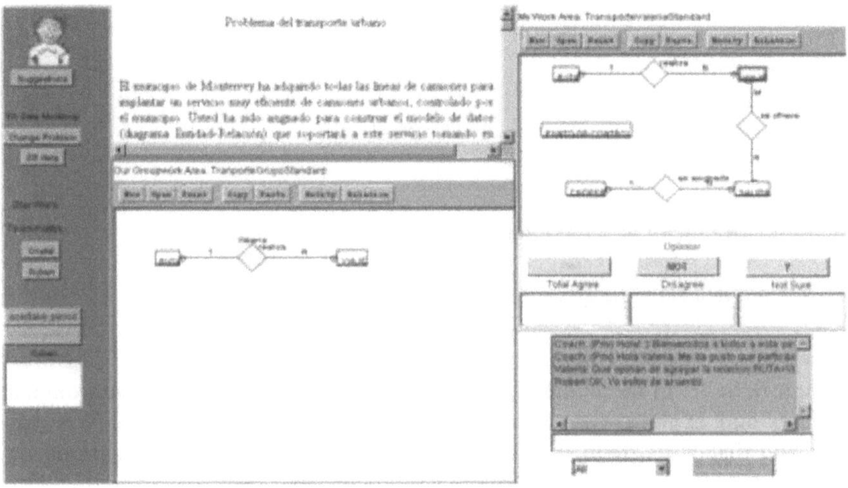

Fig. 1. COLER environment for learning ER modeling (after [5] and the related material presented at http://lilt.ics.hawaii.edu/lilt/)

4.1.1 The Context

The previous paragraph highlights several concepts and principles important for CSCL systems (italicized). Generalizing from this example and from many other CSCL systems, Figure 2 shows the context of intelligent Web-based CSCL. A group of learners typically uses a CSCL system simultaneously. The system runs on one or more *educational servers*. The learners' activities are focused on solving a problem in the CSCL system domain collaboratively. A human teacher can participate in the session too, either by merely monitoring the learners' interactions and progress in solving problems, or by taking a more active role (e.g., providing hints to the learners, suggesting modes of collaboration, discussing the evolving solution, and so on).

Given the developments and trends in artificial intelligence (AI) and in intelligent educational systems in the last decade, it is hard to imagine a truly intelligent educational system on the Web without intelligent *pedagogical agents*. Such agents provide the necessary infrastructure for knowledge and information flow between the clients and the servers in the context of Web-based education, Figure 2. They are autonomous software entities that support human learning by interacting with learners and teachers and by collaborating with other similar agents, in the context of interactive learning environments [21]. Pedagogical agents help very much in locating, browsing, selecting, arranging, integrating, and otherwise using educational material from different educational servers. As for the learners, pedagogical agents can support both collaborative and individualized learning, because multiple students and agents can interact in a shared environment. By providing rich instructional interaction with learners, they promote learner motivation and engagement, and effectively support learners' cognitive processes. Some pedagogical agents are represented by animated characters that give the learner an impression of being lifelike and believable (see the coach icon in Figure 1).

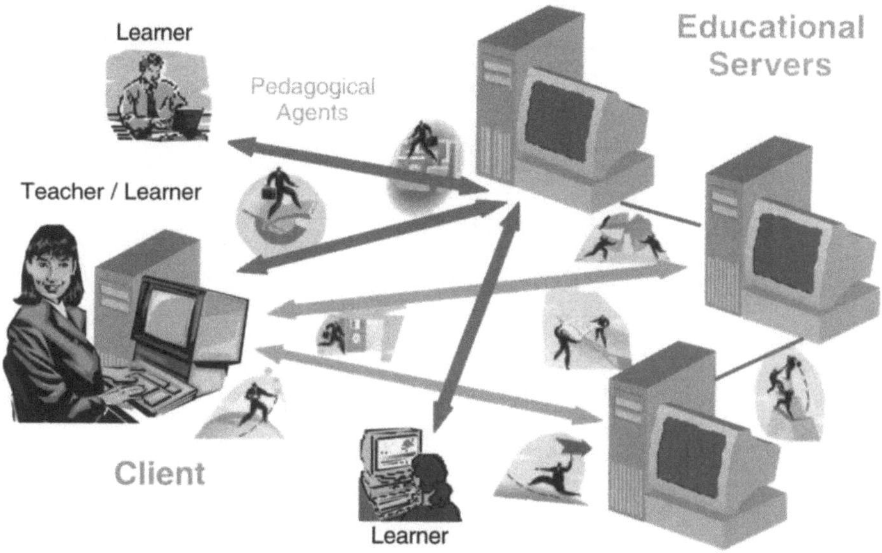

Fig. 2. The context of intelligent Web-based CSCL

Pedagogical agents access *educational content* on a server by using the services depicted in Figure 3, and the server possesses enough intelligence to arrange for *personalization* of the learning tasks it supports. In fact, from the learner's perspective the server appears to act as an intelligent tutor with both *domain* and *pedagogical* knowledge to conduct a learning session. It uses a *presentation planner* to select, prepare, and adapt the domain material to show to the student. It also gradually builds the *student model* during the session, in order to keep track of his/her actions and learning progress, detect and correct the student's errors and misconceptions, and possibly redirect the session accordingly. All the content on an educational server should be machine-understandable, machine-processable, and hence agent-ready. In order to achieve that, it is necessary to mark up the content with pointers to a number of shareable educational *ontologies*.

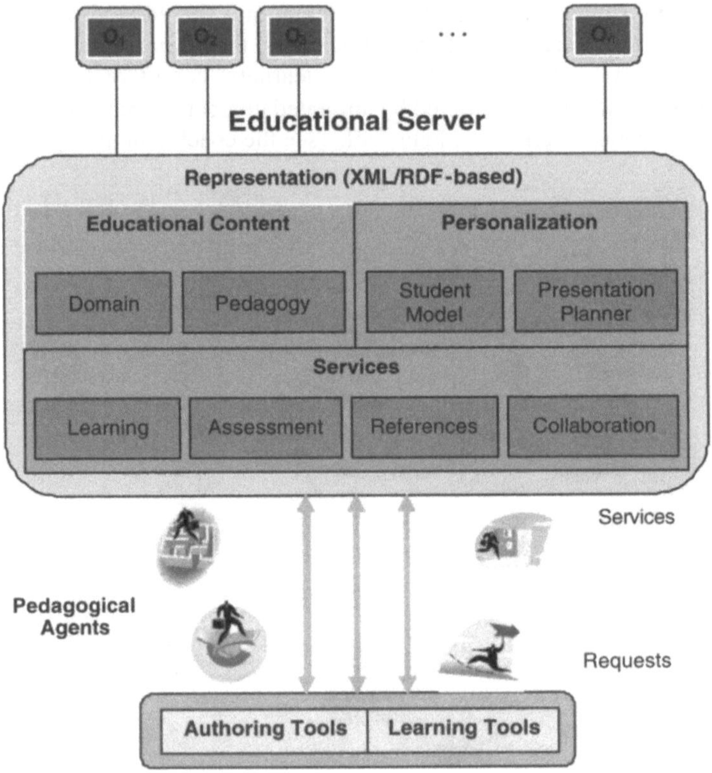

Fig. 3. Inside an educational server (O_i - ontologies)

4.1.2 Definitions and Supporting Theory

CSCL is the process related to situations in which two or more subjects build synchronously and interactively a joint solution to some problem [12], [13], [14]. Another definition says that CSCL is a coordinated, synchronous activity resulting from a continued attempt to construct and maintain a shared conception of a problem [25]. Yet another one defines CSCL as learning in which ICT is used to promote connections between one learner and other learners, between learners and tutors, and between a learning community and its learning resources (LTSN generic center, http://www.shef.ac.uk/collaborate/collaborative_elearning/index.shtml).

All such definitions stipulate both the *social context* and the *social processes* as an integral part of the learning activity. In other words, CSCL is a natural process of *social interaction* and *communication*. The goals of CSCL are three-fold:

- personal – by participating in collaborative learning, the learner attains elimination of misconceptions, development of self-regulation skills (i.e., metacognitive skills that let the learner observe and diagnose self-thinking process and

self-ability to regulate or control self-activity), and more in-depth understanding of the learning domain;
- maintaining interaction with the other learners, in order to attain the personal goal associated with the interaction; this leads to learning by self-expression (learning by expressing self-thinking process, such as self-explanation and presentation), and learning by participation (learning by participating as an apprentice in a group of more advanced learners);
- social – the goals of the learning group as a whole are setting up the situation for peer tutoring (the situation to teach each other), as well as setting up the situation for sharing cognitive or metacognitive functions with other learners (enabling the learners to express their thinking/cognitive process to other learners, to get advise from other learners, discuss the problem and the solution with the peers, and the like).

Collaborative learning is studied in many learning theories, such as Vygotsky's sociocultural theory - zone of proximal development [34], in constructivism, self-regulated learning, situated cognition, cognitive apprenticeship, cognitive flexibility theory, observational learning, distributed cognition, and many more (see [2], [12], [25], and http://tip.psychology.org/theories.html for a more comprehensive insight). Starting from these theories and applying them along with AI techniques in CSCL systems on the Web, the research community has made advances in several directions related to collaboration in learning supported by Web technologies:
- classical CSCL – this comprises setting up CSCL in Web classrooms, as well as infrastructure for CSCL in distance learning;
- learning companions – these are artificial learners, i.e. programs that help human learners learn collaboratively if they want so, even when no other peer learners are around;
- learning communities – remote learners can communicate intensively not only by solving a problem in a group, but also by sharing common themes, experiences, opinions, and knowledge on the long run;
- Web services – this general and extremely popular recent technology can be used in learning situations as well (Figure 3 depicts some possibilities);
- hybrid modes – some, or even all of the above capabilities can be supported (at least to an extent) in an intelligent Web-based CSCL system.

Since the issue of interaction is central to CSCL on the Web, it is useful to introduce the four types of interaction the learner typically meets when using such systems [7]:
- interaction with resources (such as related presentations and digital libraries);
- interaction with teachers (teachers can participate in CSCL sessions);
- interaction with peers (see the above description of the goals of CSCL);
- interaction with interface (this is the most diverse type of interaction, ranging from limited text -only interactions, to the use of specific software tools for dialogue support, based on dialogue interaction models, to interaction with pedagogical agents (see the dedicated section later in this chapter).

One should differentiate between *cooperative* and *collaborative* learning [16]. In cooperative learning, the learning task is split in advance into sub-tasks that the partners solve independently. The learning is more directive and closely controlled by the teacher. On the other hand, collaborative learning is based on the idea of building a consensus through cooperation among the group members; it is more student-centered than cooperative learning.

4.1.3 Summary – The CSCL Model

Each effective CSCL system is based on a fairly general model, integrating the following four important issues [27]:

- indicators of effective collaborative learning;
- strategies for promoting effective peer interaction;
- technology (tools) to support the strategies;
- a set of criteria for evaluating the system.

CSCL system should recognize and target group interaction problem areas. It should take actions to help students collaborate more effectively with their peers, improving individual and group learning.

4.2 Architectural Issues

Generally, software architecture is a specification and design of the overall structure of a software system. There are a number of issues related to software architecture, such as gross system organization, the choice of architectural style, global control structures, protocols for communication, synchronization and data access, selection among design alternatives, selection of design elements and definition of their interfaces and collaboration, assignment of functionality to design elements, composition of design elements, physical distribution of system components, scaling, and performance.

4.2.1 CSCL Architectural Styles

A software-architecture issue of particular interest for CSCL system is that of architectural style. Architectural style is related to a family of systems organized in a similar way. Most of current Web-based CSCL systems are built using one of the following four architectural styles: centralized, replicated, distributed, and hybrid (between distributed and replicated) [29]. These four styles are easily understood having in mind the well-known Model-View-Controller paradigm often used in software design (see, for example, [15] for details). In the *centralized architecture*, the CSCL application is running on a single host machine, whereas copies of the GUI events coming from View and Control components on the server are sent to clients (the machines used by the learners). Tight coupling between different users (learners) based on strict copies of the GUI events can be relaxed to independent scrolling, editing, or moving of objects on the screen. However, due to a number of GUI events that have to be copied to each client machine, the use of network bandwidth in this architecture is rather inefficient. In the *replicated architecture*, application is installed and run on each client machine and synchronization between the clients is enforced when necessary. This ensures for better use of the network bandwidth (only Controller events are transmitted for synchronization purposes), but application must be designed with collaboration in mind.

In a *distributed architecture*, the Model component runs on a shared server (e.g., a database), and each client has its own View and Controller. This flexible Model-level coupling enables having different Views on different client machines to suite different learning styles. Only Model update events sent over the network. This is good in terms of efficiency, but can also be a disadvantage since the entire system relies heavily on the network. Hence the current view is that the most suitable architectural style for CSCL applications is *hybrid architecture*. As in the distributed architectural style, hybrid architectures comprise model-level synchronization via a persistent model maintained on a dedicated server machine. However, the model is also replicated in the client machines, allowing applications to run standalone with their own models, saving state to the local file system. Model updating between active clients is done through the persistent store (persistent model).

4.2.2 Open Distributed Learning Environments

There is another trend in software architectures for CSCL – *open distributed learning environments* [22], [20], [23]. The idea here is that learning environments and support systems are not conceived as self-containing, but as embedded in realistic social and organizational environments suitable for group learning. This idea is illustrated in Figure 4. In an integrated Web-based classroom, the learners solving a problem collaboratively can communicate face-to-face or electronically (through the local network). They can also communicate with the teacher, asking for help, hints, opinion, explanations and other guidance. However, different Web classrooms can be interconnected among themselves letting the learners communicate with the peers and teachers not physically present in the same classroom, but logged onto the same network/application. Moreover, the CSCL application can link the learners with other relevant learning resources or remote peers and tutors through the Internet.

Technology supporting Web classrooms includes the server(s) running the learning application, local student computers (desktop, notebooks, tablets, ..., all locally networked and with access to the Internet), a Live Board for presentation-based learning, as well as a number of accessories, such as light pens, electronic markers, signal buttons, etc.

Note, however, that Web classrooms themselves are not just local computer networks facilitating computer-based learning and substituting traditional teacher-moderated classrooms. Web classrooms can be classrooms with physically present learners, as well as virtual classrooms with remote interaction. In both cases, they are driven by CSCL software applications enabling intensive electronic communication and collaboration between the learners and teachers during the problem-solving process. Their primary goal is to *support* collaboration and make the learning *more effective*, hence they are different from simple networks of learning resources. Specific CSCL software components can include shared workspace tools for different domains, general "discussion board", tools for taking and collecting individual notes, electronic worksheets that can be distributed and collected by the teacher, and intelligent monitoring tools that can give both individual feedback and information for the teacher [23].

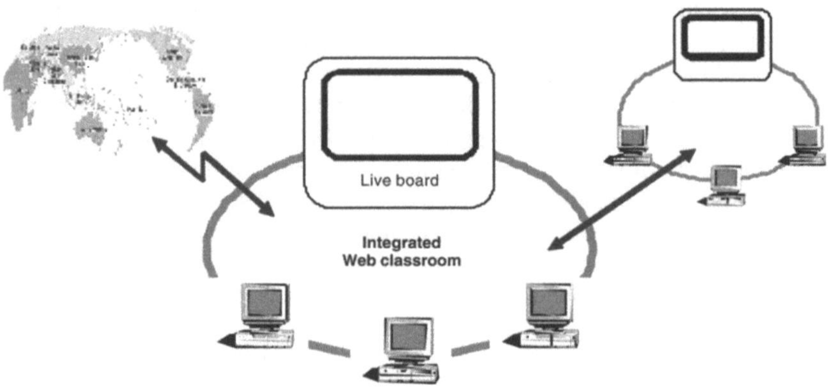

Fig. 4. Open computer-integrated Web classrooms (after the idea from http://www.collide.info/)

4.2.3 An Example

In order to illustrate the architecture of Web classrooms and CSCL systems in more details, Figure 5 shows an example. The architecture shown is used in the intelligent Web classroom running Code Tutor, the system for learning radio communications, Figure 6 [31]. The details in Figure 5 specify the technology used to establish the Web classroom. It is an example of hybrid architecture, in which learner models (Student models), and domain knowledge (Expert module) are stored persistently on a server, but are replicated on both the teacher and the students sides. The application has the teacher component and the student component. The two components allow for different views and different control authorities over parts of the application (see the details specified for GUIs at the two sides in Figure 5). It is important to note that this Web classroom and the Code Tutor application are based on well-established and widely used technologies:

- CLIPS, a tool for building expert systems is used to generate knowledge base files;
- Java-based expert system shell Jess is used to interpret these files;
- Students communicate with the system through a standard Web browser;
- Java Servlet tehnology is used to implement the system's interactions with the students;

- Apache server is used to store static HTML pages;

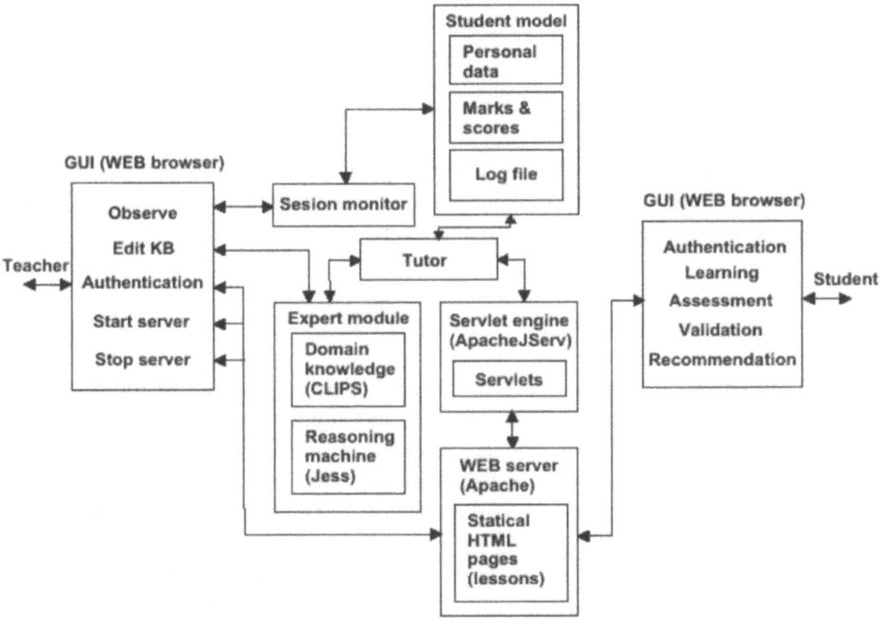

Fig. 5. An example architecture of an intelligent Web classroom (after [31])

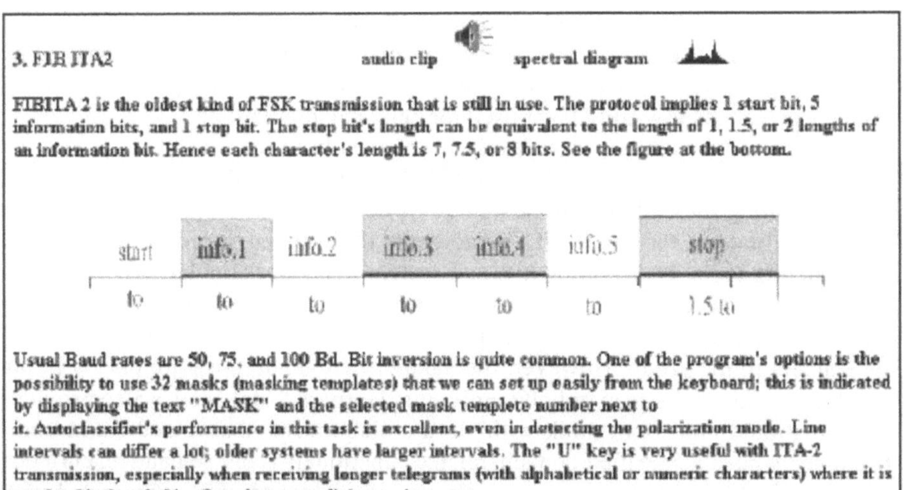

Fig. 6. A screenshot from Code Tutor

- Apache JServ is used to interpret the servlets;

- XML technology is used to generate files that Code Tutor uses to provide recommendations to the students.

4.3 Design Issues

There are several current trends in designing intelligent Web-based CSCL systems. Some of them are more CSCL-specific. Others stem from current software technologies and are common for other software systems as well.

4.3.1 User Interface Design

It is quite understood that the learning process is more effective if the user interface is designed to be intuitive, easy-to-use, and supportive in terms of the learners' cognitive processes. With CSCL systems, additional flexibility is required. The learners have to work collaboratively in a shared workspace environment, but also use private workspaces for their own work. Moreover, since work/learning happens in small groups, the interface should ideally support the group working in one environment, or in synchronous shared environments. It also must support sharing of results, i.e. exchanging settings and data between the groups and group members, as well as demonstrating the group's outcomes or conclusions. A suitable way to do it is by using a public workspace.

This division of learning/work into shared and private workspaces leads to the idea of workspaces that can contain a number of *transparent layers* [20], [23]. The layers can have "solid" objects (synchronizeable visual representations), e.g., handwriting strokes or images. Also, the layers can be private or shared. For example, a private handwriting layer used for personal annotations.

The idea of using transparent layers in the design of user interface is best exemplified in the intelligent Web-based CSCL called Cool Modes (COllaborative Open Learning, MOdelling and DEsigning System) [23]. The system supports the Model Facilitated Learning (MFL) paradigm in different engineering domains, using modeling tools, construction kits and system dynamics simulations. The focus of the learning process is on the transformation of a concrete problem into an adequate model. The shared workspace, Figure 7, is public and looks the same to all the learners in the group. However, the handwritten annotations are placed in private layers and can be seen only by individual learners. Cool Modes also provides "computational objects to think with" in a collaborative, distributed framework. The objects have a specified domain-related functionality and semantics, enriched with rules and interpretation patterns. Technically, Cool Modes is integrated with visual modeling languages and has a set of domain-specific palettes of such objects (see the palette on the right-hand side of Figure 7). The palettes are defined externally to encapsulate domain-dependent semantics and are simply plugged-in the system when needed. Currently, the system has palette support for modeling stochastic processes, system dynamics, Petri nets, and other engineering tools, as well as for learning Java.

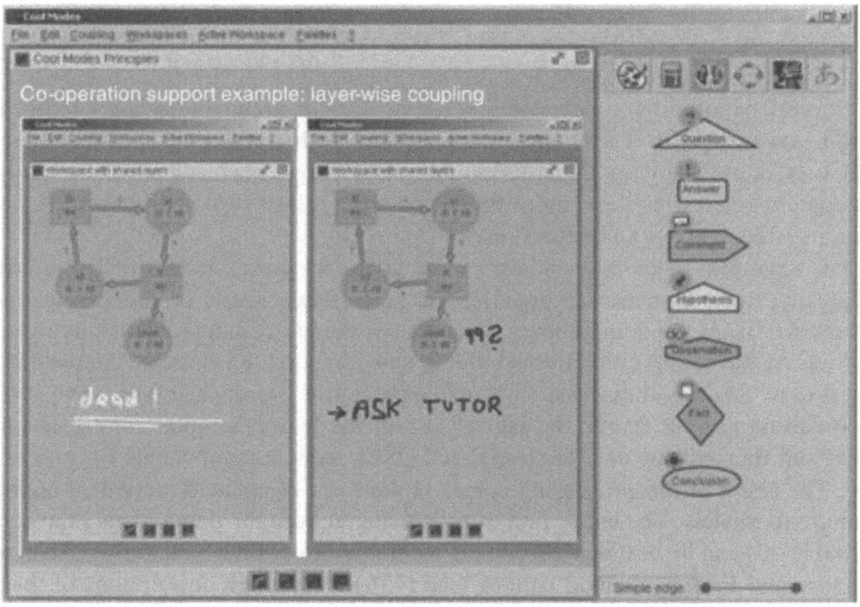

Fig. 7. A screenshot from Cool Modes (after [23])

The teacher has to prepare the initial problem descriptions and the specific settings for the learners using the corresponding visually orientated microworld (the corresponding palette). Then the learners solve the problems through Cool Modes using the same palette, constructing the model of the solution themselves as a group, or in cooperation with the teacher. The teacher can help the learners during the session using annotation elements and free hand input.

4.3.2 Software Patterns for CSCL

In software engineering, patterns are attempts to describe successful solutions to common software problems [15], [26]. Software patterns reflect common conceptual structures of these solutions, and can be applied over and over again when analyzing, designing, and producing applications in a particular context. Patterns emerge from successful solutions to recurring problems and with the knowledge of patterns it is not necessary for a developer to solve every software problem again and again from scratch; instead of that, he can profit from the experience other software engineers had with similar problems.

There are also patterns in the broad area of education. *Educational design patterns* are patterns of educational processes and activities. Likewise, *instructional patterns* are common procedures for achieving desired instructional goals – there are many kinds of patterns in the way learners learn and in the way teachers teach. For example, *spiral instructional pattern* expresses the typical Doing-Assessment-Adjustment cycle, and *coaching strategy pattern* represents the Decide-Implement-Assess procedure in coaching. See [10] for a more elaborate discussion on patterns in education and in learning environments.

Several patterns have been discovered so far for architectures of intelligent CSCL systems [10]. They can be applied for both traditional and Web-based CSCL systems. Applying them in designing CSCL systems results in a shorter development and testing time, more stable and more robust system organization, and easier system updates and extensions.

For example, there is a pattern in the gross organization of CSCL systems. Typically, such systems are organized to reflect two levels of processing raw data/material. At the domain level, each learner is involved in problem solving activities. At the conversational level, the system supports the learners' cooperation, division of labor, and discussion. Raw data/material is collected through specialized editors at both levels; the editors accept the learner's input and create raw data, and transmit the data to specialized CSCL system components for processing. The result of this processing comes in the form of updated individual learner and group models. The entire line of processing at both the domain and conversational levels can be best designed using a combination of the well known Abstract Factory and Factory method pattern (see [15] for thorough descriptions of these patterns).

As another example, recall that learning companion systems enable collaborative learning with artificial learners, or co-learners. Such co-learners can take different roles, such as:

- *learning companion*, which learns to perform the same learning task as the student (human learner), at about the same level, and can exchange ideas with the student when presented the same learning material;
- *troublemaker*, which tries to disturb the student by proposing solutions that are at times correct, but are wrong at other times, thus challenging the student's self-confidence in learning;
- several *reciprocal tutoring* roles;
- roles that aim to stimulate collaboration and discussion within learning communities, such as *observer*, *diagnostician* or *mediator*.

It is important to note that *architecturally*, all CSCL systems involving a co-learner have much in common regardless of the role the co-learner takes. In fact, we can talk about the *Co-Learner pattern*, represented in Figure 8. The classic 3-agents triad – *Tutor-Student-Co-Learner* – is shown in Figure 8a with more details than the literature on co-learners usually offers, due to the fact that all pattern diagrams have to show both the participants and their communication paths clearly. Hence that part of the figure stresses the "who communicates with whom" and "what knowledge and data are involved" issues. Moreover, since this is an architectural view, it is necessary to show details more-or-less irrelevant for instructional aspects of co-learner systems. For example, the *System* component acts as a supervisor and performs all the control and scheduling of activities of the three major agents. It is shown grayed, though, since it doesn't contribute to essentially to the major knowledge and information flows. Furthermore, Figure 8a clearly indicates what kinds of knowledge and data each agent needs. Put this way, it turns out that the Co-Learner pattern belongs, in fact, to *blackboard architectures* [3]. All knowledge and data are on the blackboard, but usually only the *Tutor* agent accesses all of them. *Student* and *Co-Learner* normally have access only to the *Learning Task* part of the blackboard (thick data-flow lines). Variants are discussed in later paragraph (dashed data-flow lines).

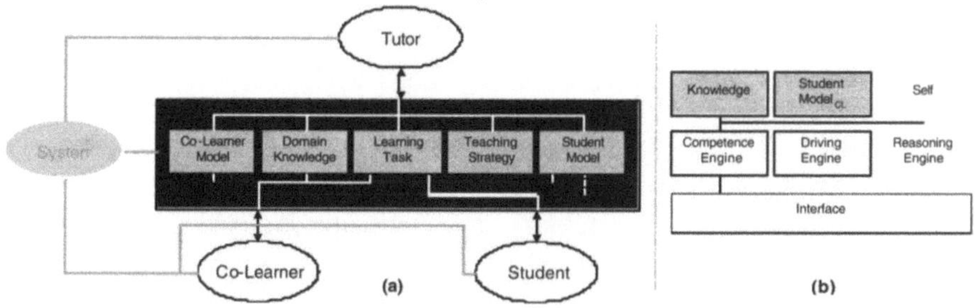

Fig. 8. Co-Learner pattern a) communication paths b) inside the Co-Learner

Figure 8b shows a fairly generalized version of the Co-Learner agent itself. Depending on the role, it can have more or less of its own *Knowledge*, both domain and pedagogical. If Co-Learner has little knowledge, it is a novice learner; if its knowledge is comparable to *Tutor*'s, it is an expert. That knowledge can grow over time for teachable agents, and in all reciprocal-tutoring cases when Co-Learner switches the teaching and learning roles with another agent in the system. For troublemakers, that knowledge can include details of the learning by disturbing strategy (although it is possible, in principle, to access it on the blackboard as well). Co-Learner constructs and maintains its own model of the human learner's knowledge and progress, *Student Model* $_{CL}$, which is generally different from the *Student Model* built by *Tutor*. Also, the agent's own internal state (attribute values, learning status, the level of independence, motivation, personality characteristics, the corresponding animated character (if any)) is stored in *Self*. Systems with multiple co-learners can have Co-Learner agents with different *Self* characteristics.

4.3.3 Integration of Multiple AI Techniques

When designing intelligent Web-based CSCL systems, developers often use multiple AI techniques. There are many possible combinations. As an illustration, consider the Web-based CSCL system Lin2k–C/R-FIS that combines fuzzy logic and expert systems [18]. A screenshot from one of its applications – collaborative composition of simple musical scores – is shown in Figure 9. Lin2k–C/R-FIS is designed to enforce *collaborative activity equilibrium* when used in collaborative learning situations involving two peers – the contributions of both learners to the joint solution should be more or less equal. Essentially, Lin2k–C/R-FIS is an aggregation of two AI systems:

- Lin2k – a Web-based tool for asynchronous, written collaboration between two peers;
- C/R-FIS – an expert system that automatically evaluates the collaborative activity (contribution) of the two peers; the evaluation results are used for adaptive support to peers during their collaboration.

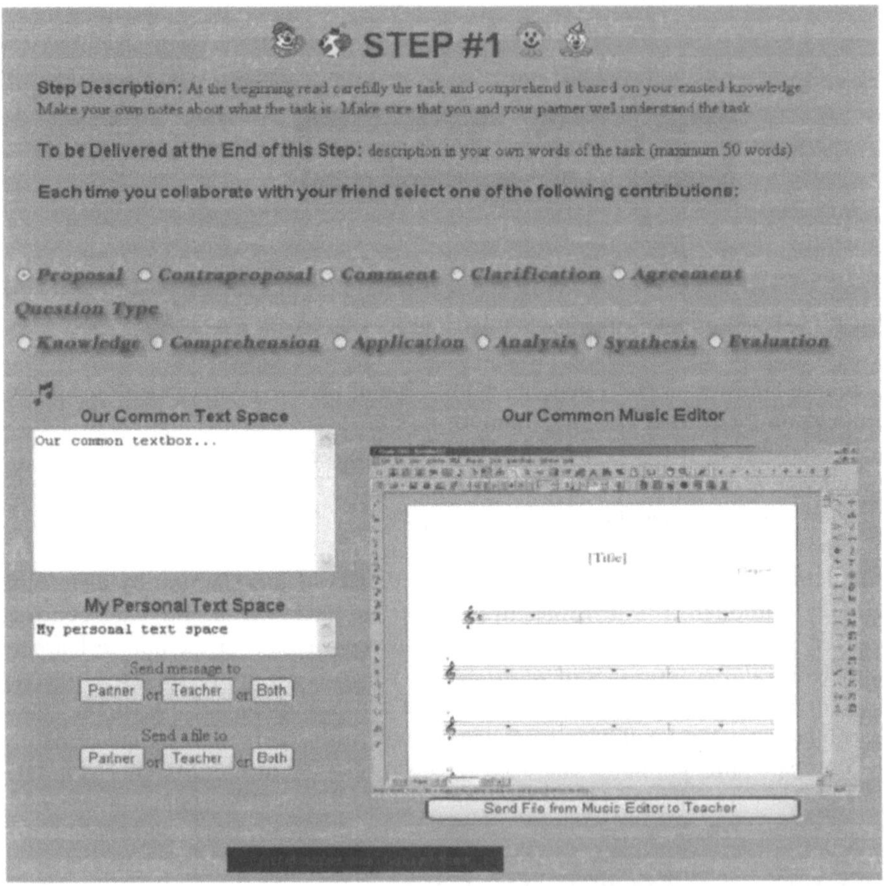

Fig. 9. A screenshot from the Lin2k–C/R-FIS system (after [Hadjileontiadou et al. 2003])

The expert knowledge in C/R-FIS is expressed through fuzzy rules of the following form:

IF **initiative is satisfactory AND creativity is low**
THEN **collaboration is low**

IF **work is low OR initiative is low OR questions is low**
THEN **argumentation is low**

The system provides an easy-to-use interface for experts in cognitive psychology to express such rules, without worrying much about their internal representation in fuzzy logic.

On the other hand, Lin2k maintains logs recording the learners' activity at the task level (by mouse clicks) and written dialogue pertaining to the coordination of collaborative activity. Interfaces display discussion threads at both levels, to facilitate their development.

Evaluation of each peer's collaborative activity is based on a weighted sum of four variables: initiative, creativity, elaboration, and disagreement. The weights are defined empirically at both the task and collaboration levels. The values of the variables are computed from contributions during collaborative activity. There are seven types of contribution: proposal, counterproposal, comment, clarification, agreement, low-level questions, and high-level questions. These are used to express fuzzy rules in C/R-FIS. Also, each peer selects explicitly the type of contribution each time he/she contributes to the solution, hence the values of the four variables are regularly updated. From the elaboration of each peer's collaborative activity, the system is able to enforce a balanced collaborative activity of the peers.

4.3.4 Software Components and Web Services

A software component is a physical packaging of executable software with a well-defined and published interface [19]. Component software addresses the general problem of designing systems from application elements that were constructed independently by different developers using different languages, tools, and computing platforms. [1]. Software components are such of independent production, acquisition, and deployment that interact to form a functional system [30].

Several designers of Web-based CSCL systems have used component software technology. Examples of such systems are Belvedere (http://lilt.ics.hawaii.edu/lilt/software/belvedere/index.html), and E-Slate (http://e-slate.cti.gr/). These systems treat an educational software component as a piece of a larger system that can be plugged in, taken out, and used together with other components to contribute to the global system's behaviour. For example, E-Slate components are provided as a kit of pre-fabricated, interoperable computational objects. Educational software is created with these components in a rapid prototyping manner with minimal effort and resources. Components encompass domain-specific knowledge, playing the role of high-level building blocks in the creative hands of authors.

In order for educational software components to be used in system development effectively, it is necessary to have a software infrastructure to support the "plug-and-play" design philosophy of component-based systems. This infrastructure is called *component framework*. It is a low-level software basis for developing educational software components in and categorizing them within a higher-level framework for developing educational software. An example of such a higher-level framework is GET-BITS (see [11] for details). Note that when one builds knowledge and reasoning capabilities into educational software components, they become pedagogical agents.

Web services are a step beyond software components. They are activities allowing both end users and, under appropriate circumstances, software agents to invoke them directly [24]. In the traditional Web model, users follow hypertext links manually. In the Web services model – and Figures 2 and 3 comprise using *that* model – they invoke tasks that facilitate some useful activity (e.g., meaningful content-based discovery of learning material, fusion of similar educational material from multiple sites, or commercial activities such as course advertising and registration for distance learning). Technically, Web services are autonomous, platform-independent computational elements that can be described, published, discovered, orchestrated, and programmed using XML artifacts for the purpose of developing massively distributed interoperable applications. Platform-neutral and self-describing nature of Web services and particularly their ability to automate collaboration between Web applications make them more than just software components.

Note, however, that in collaborative learning situations on the Web the idea is to employ *intelligent* Web services – to go beyond XML/RDF infrastructure of Web pages, to explore Web services that can be enabled by intelligent systems technology. The learners, the teachers, and the authors of Web-based CSCL systems alike should be able to see the Web as if it was turned into a huge collection of educational resources, each with a well defined interface for invoking its services [33]. Automatic discovery, invocation and composition of educational Web services on behalf of the learners should be the job of pedagogical agents.

Although using educational Web services in CSCL systems is still not widespread, the appropriate technology is ready. An example of a *service-oriented architecture* of Web-based educational servers, elaborated after [33], is shown in Figure 10 (recall the internal structure of educational servers, shown in Figure 3). Educational services should advertise themselves in a registry/directory, allowing the learners' agents to query the registry for service details and interact with the service using those details. Such a directory is an information pool pertaining to different educational services, dynamically organized, but highly structured (e.g., as a tree, or as a table/database). The underlying assumption is that at each point in time the directory lists those services that are ready to be invoked by the learner or by a group of learners; the services are supposed to advertise their readiness and availability to the directory. Hence a pedagogical agent can find out about the available services by looking up the directory. Then it can decide whether to automatically invoke a suitable service on the learner's behalf, or merely to suggest the learner to interact with the service directly.

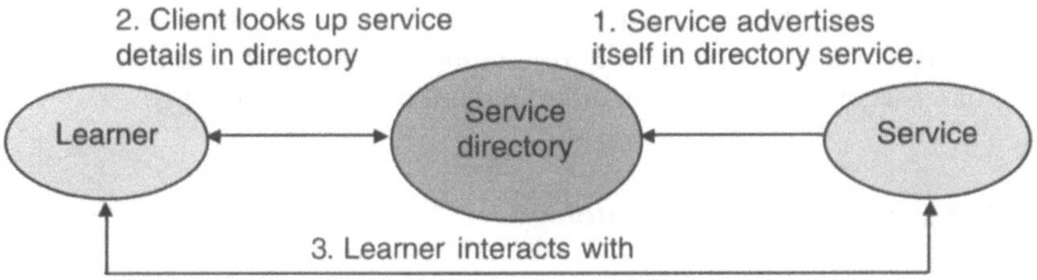

Fig. 10. Service-oriented architecture of educational Web servers

How exactly educational services can help in CSCL situations? To answer this question, we should refer again to Figures 2 and 3. Suppose groups of learners are created dynamically and opportunis tically, when needed. A learner wishing to join a group can issue to her pedagogical agent a request like "Find group", or even "Create group". Likewise, a group of learners may need a teacher to join the group, issuing a request like "Get teacher". Such requests are processed by pedagogical agents by looking up an educational service directory. As a result, the agents will easily locate the educational servers supporting collaboration services (Figure 3) that advertised their services in the directory, and then automatically invoke the requested service. Specific collaboration services on educational servers may include group formation and matching, class monitoring, discussion support, chat, groupware, video/audio, and the like.

4.4 CSCL and the Semantic Web

Semantic Web (http://www.semanticWeb.org/) is the new-generation Web that makes possible to express information in a precise, machine-interpretable form, ready for software agents to process, share, and reuse it, as well as to understand what the terms describing the data mean. It enables Web-based applications to interoperate both on the syntactic and semantic level. The key components of the Semantic Web technology are [24]:

- its unifying data model (currently, the most used one is RDF (Resource Description Framework, http://www.w3.org/RDF/));
- ontologies of standardized terminology that represent domain theories; each ontology is a set of knowledge terms, including the vocabulary, the semantic interconnections, and some simple rules of inference and logic for some particular topic;
- languages based on RDF, such as OWL (Web Ontology Language, http://www.w3.org/TR/2003/WD-owl-guide-20030331/), for representing ontologies and for marking up Web resources.

Returning to Figure 3, note that educational Web servers are supposed to point to a number of ontologies. Ideally, creation of educational Web contents with ontological annotation should be supported by ontology-driven authoring tools and class hierarchies based on a number of underlying ontologies. Teaching and learning contents of Web-based educational applications can then be presented, edited, modified, and mixed *consistently*. Furthermore, ontologies should be linked to libraries of terms, and interlinked in order to reuse or change terms [8].

4.4.1 Collaborative Learning Ontology

At the moment, educational ontologies are still scarce – developing ontologies of high usability is anything but easy, and the Semantic Web is around for just a couple of years. Still, CSCL community has already ventured in developing CSCL ontologies. Pioneering but extremely important work in this direction came from the Mizoguchi Lab, Osaka University, Japan. They have made considerable effort towards developing Collaborative Learning Ontology (CLO) [28]. Although still not widely used, CLO clarifies the concepts of a collaborative learning group, and the relations among the concepts. It answers general questions like:

- What kinds of groups exist in collaborative learning?
- Who is suitable for attaining the group?
- What roles should be assigned to the members?
- What is the learning goal of the whole group?

Reflecting on these questions leads to the illustration in Figure 11. Collaborative learning is a learning process to which an individual learner can switch from learning individually. To this end, an important concept in the CLO is *trigger* – detection of an opportunity for a learner to shift from individual learning mode to collaborative mode. Such an opportunity may arise, for example, when an individual learner comes to an *impasse*, or wants a *review* of his solution to a problem, and so on. The highlighted concepts are also important in the CLO. Furthermore, CLO defines *learning group* as the collection of learners with the same learning goal, as well as related concepts like *group-type*, *learner's role in the group*, *learning goal* (learners' goal for from collaboration perspective), *group-performance*, *learning efficiency*, *learning scenario* (outline of the learning subject, process, and material), *learning topic*, and many more. Some of these concepts are depicted in Figure 11.

The upper part of Figure 11 indicates the need for an intelligent CSCL application to use another ontology in addition to CLO – the *negotiation ontology*. Some of its concepts are shown in the upper left corner of the figure. The idea is that the decision on the shift from individual learning mode to collaborative one should be made carefully, i.e. through a *negotiation process*. The learning environment should be capable of *devising* the collaborative learning mode, based on *observation* of the learner's progress in individual learning mode, *investigation* of other learning possibilities, *compromise* between the learning styles, or even some *persuasion* to shift to collaborative learning. All such concepts should be defined explicitly in the negotiation ontology. Based on such considerations, the system needs to manipulate different *negotiation objects* (proposal, justification, conflict, criticism,...) and generate different *negotiation events* (send, receive, reach-agreement), resulting in invoking appropriate *negotiation performatives* (call-for-participant, reply, support, give-opinion,...). for example, if the system decides to *devise* shift to collaborative learning, it can generate and *send* a *proposal* about it to the learner, invoking the *call-for-participant* performative. Again, all these concepts and actions should be defined explicitly in the negotiation ontology.

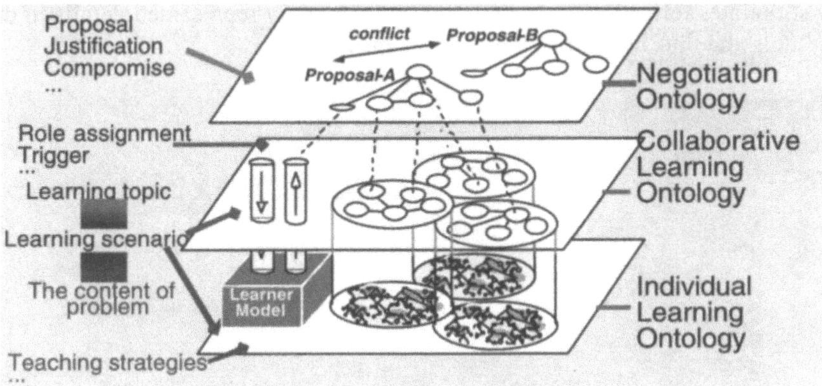

Fig. 11. The context of the Collaborative Learning Ontology (after [28])

4.4.2 Learning about Petri Nets Collaboratively

Another example of how ontologies can effectively support collaborative learning on the Semantic Web is related to the Petri net ontology and P3 tool for Petri nets development, analysis and simulation [17]. Petri nets are known to be a useful tool for modeling and simulation in many engineering domains. There are different tools for Petri net development on the Internet, many of them free to download. Although most of them do support most of the Petri net concepts and multiple Petri net dialects, there are also minor differences among all these tools. That was the fact that we used for setting up a collaborative learning environment for learning Petri nets. The environment is actively used in our course on Computer Architectures, where Petri nets can be used effectively to simulate different processes in computers.

Essentially, the environment is a loosely coupled aggregation of different Petri net analysis and development tools, such as P3 (the tool developed in our lab; it can be downloaded from http://www15.brinkster.com/p3net). When the same Petri net is developed using two different tools, the resulting models will in most cases differ in some details, Figure 12. These minor differences can be very useful for learning details about Petri nets. When two learners are given a problem in the area of Petri net modeling and different tools to support the modeling activity, the resulting models will be constrained by the corresponding tool capabilities. However, since current Petri net modeling tools usually support export to and import from XML formats, it is rather straightforward to convert from one format to another using XSL transtormations (XSLT). Petri net ontology that we developed in RDFS and in OWL is used to interpret the differences, i.e. to validate the resulting models before and after XSLT. Thus, one learner can partially develop a Petri net model using one software tool and export/convert it to another tool format. Another learner, using the other tool, can add other details. Some of them may be enabled only by the other tool. Then he can export the model back to the first format. This way, the model grows as a result of a collaborative activity and simultaneously stimulates reflection on the missing or differently represented details in different tools and thus intensifies the learning process.

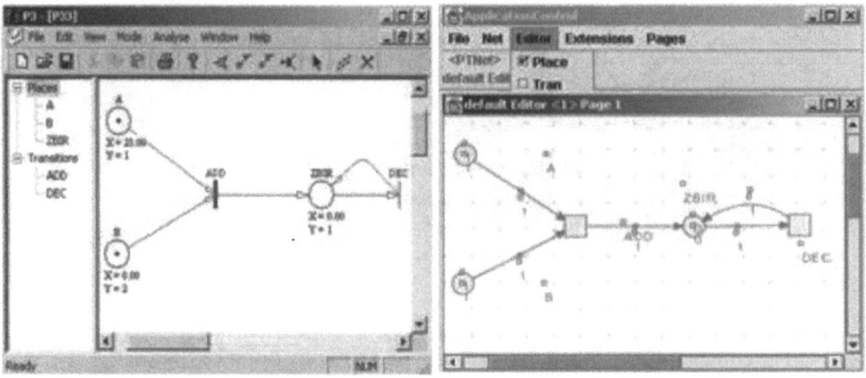

Fig. 12. Learning about Petri nets collaboratively using different tools

4.5 Agent Support for CSCL

In the context of intelligent Web-based CSCL, Figure 2, pedagogical agents are important intelligent software entities that support a group of learners and teachers in their interaction with the learning objects on the Web, educational servers, and Web-based learning environments. Pedagogical agents can take different roles and can help the learners in many ways.

4.5.1 Agent roles in CSCL Environments

There are two broad categories of agent roles in CSCL – coordination managers and coordination facilitators [35]. *Coordination managers* mediate administrative aspects of collaboration. For example, agents can mediate access to and control of shared workspacesIn case of simultaneous or conflicting requests for control of shared workspace, coming from multiple learners, such coordination managers can take care of scheduling and granting access/control to individual learners. Moreover, communication between the learners, conferencing tools, and different shared panels (see Figures 1 and 9) can be also controlled by coordination managers. Finally, coordination managers can be in charge of different notification, booking, monitoring, and information mining tasks for both individual learners and groups.

Coordination facilitators support students involved in collaborative learning activities by mediating processes such as development of shared goals and visions, planning, searching, sorting, and sharing of information. They can also support team building, group decision-making, joint thinking, negotiation, etc.

Examples of concrete roles that can be taken by different pedagogical agents are:

- coordination managers – schedule manager, synchronization manager, competition manager (for collaborative learning tasks involving competition among the peer learners), competition monitor (-II-);
- coordination facilitators – team building agents, group decision making agents, decision-making tutor agent, team assignment agent.

4.5.2 Opportunistic Group Formation

As an illustration of how pedagogical agents can take different roles and perform different tasks in Web-based CSCL settings, consider the task of opportunistic group formation (OGF) [28]. If for any reason a learner wants to participate in collaborative learning on the Web, learning efficiency depends on joining an appropriate learning group. Hence the question "How to form a group" is important. OGF is a framework that enables pedagogical agents to initiate, carry out, and manage the process of creating a learning group when necessary and conducting the learner's participation to the group. Agents in OGF support individual learning, propose shifting to collaborative learning, and negotiate to form a group of learners with appropriate role assignment, based on the learners' information from individual learning.

In OGF, collaborative learning group is formed dynamically. A learner is supposed to use an intelligent, agent-enabled Web-based learning environment. When an agent detects a situation for the learner to shift from individual to collaborative learning mode (a "trigger", such as an impasse or a need for review), it negotiates with other agents to form a group. Each group member is assigned a reasonable learning goal and a social role. These are consistent with the goal for the whole group.

Technically, the above scenario happens as in Figure 13. When the monitoring agent detects the trigger for shifting the learner to collaborative learning mode, it broadcasts the requests on behalf of its user. It means that the agent negotiates with other pedagogical agents it knows of about the possibility to form a learning group. If they conclude the possibility is viable, the group is created. The agents then assign appropriate roles to the group members, such as "participant", "leader", "helper", and so on. Note that it is necessary to implement the negotiation in OGF according to a negotiation process model based on learning theories. There is always a trade-off between personal and social learning aspects, hence justification is necessary to form a group. In some cases, individual learning may be preferred to collaborative.

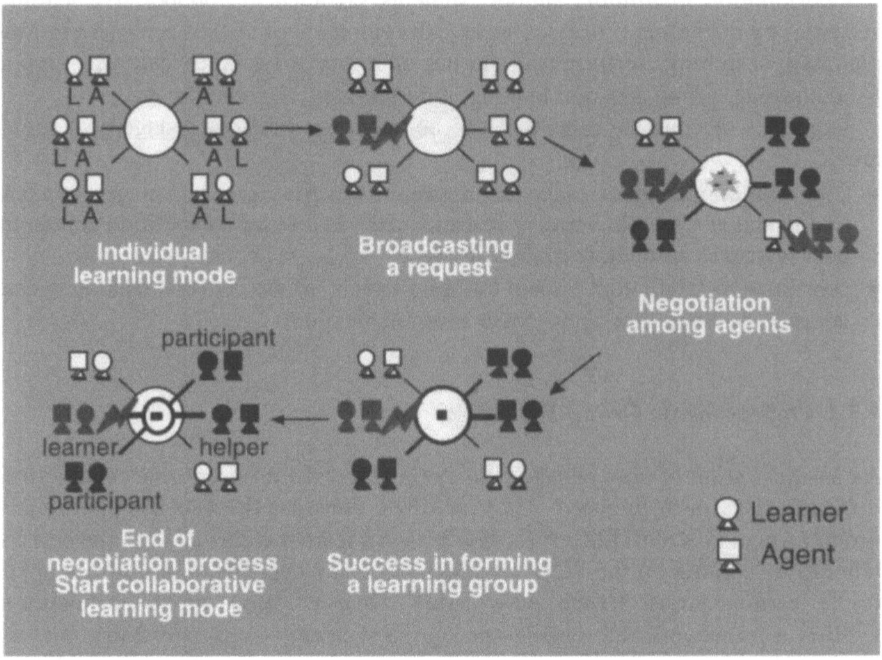

Fig. 13. The principle of opportunistic group formation (after [28])

4.5.3 Peer Help Environments

Another interesting application of pedagogical agents in CSCL is that of establishing face-to-face collaboration between human learners. I-Help system, developed at the University of Saskatchewan, Canada [32], is a typical example. It is a peer help environment capable of locating appropriate human and electronic resources for learners on the Web. It provides support for learner collaboration and peer help based on a community of pedagogical agents that operate in the way similar to the OGF framework, but with a different goal – the agents help the learners find the right person(s) to talk to face-to-face when they need help.

In I-Help environment, each learner has his/her personal agent. Whenever a learner needs help in completing a learning task, he/she contacts the agent. The agent may respond in several ways. It may try to provide help to the learner in the form of a reminder of the learning resources the learner has already used (and may have overlooked them in this case); these are stored as historical information in the learner's personal log files. It may also try to direct the learner to Web resources relevant for his/her learning topic. The most interesting case is when the learner needs peer help from another learner. In that case, his/her agent contacts a facilitator agent called Matchmaker. This agent knows about personal agents of many other learners. Typically, I-Help system will be installed on a university campus and will be used by a number of students, so Matchmaker will know about their agents from accessing appropriate databases. Upon request for peer from a personal agent, Matchmaker will contact the other personal agents and ask them to forward the request to their masters. The other students will receive the message from their agents and may then decide to contact the learner who needs help directly. There is a number of heuristics that Matchmaker can use when deciding on what other personal agents to forward the request for help, such as GPA of other students, the results the other students got when taking the exams in relevant courses, the other students' records on providing peer help to their colleagues in the past, and so on.

4.6 CSCL and Web Mining

Web mining is the process of discovering potentially useful and previously unknown information and knowledge from Web data [6]. It encompasses tasks such as automatic resource discovery, automatic extraction and pre-processing of desired data from Web documents, discovery of common patterns across different Web sites, and validation and/or interpretation of discovered patterns [4]. Figure 14 shows the most important categories of Web mining.

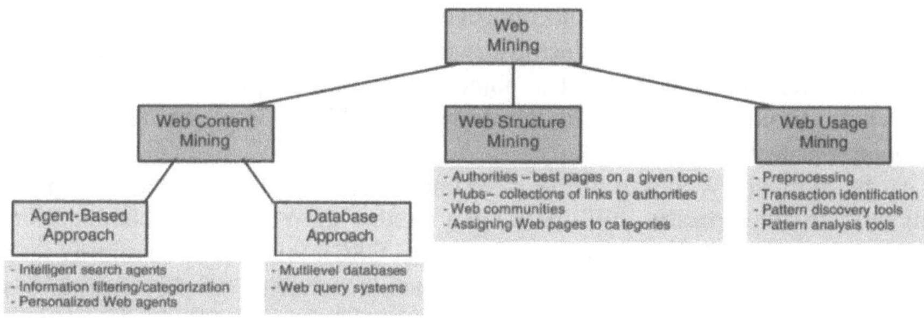

Fig. 14. Web mining categories (after [6] and [4])

All categories of Web mining are of interest for CSCL. Personalized, ontology-enabled pedagogical agents can be deployed to continuously go *Web content mining* to collect globally distributed content and knowledge from the Web (large Web data repositories such as documents, logs, and services) and organize it into educational Web servers. The collected data can then be incorporated with locally operational knowledge/databases to provide groups of learners with centralized, adaptable, intelligent Web services.

Further possibilities stem from deploying *Web structure mining* on an educational server's side. Note that a number of CSCL tasks may take a number of sessions to be completed, and the sessions can be scheduled over a long period of time. During that time, the server can continuously mine the Web for refreshing the information in its database about the availability of external educational services, ranking the most authoritative Web pages and services on a given topic, or (re)organizing its local hub of links to such external pages and services. Thus the learners can always get up-to-date information on a given topic from the educational server.

Probably the most attractive Web mining category for Web-based CSCL is *Web usage mining*, which is related to discovering typical patterns of how the users browse, access, and invoke Web pages and services. All user activity is stored in log files on Web servers, hence such files represent a rich source of data for automatic pattern discovery using suitable data mining techniques. In Web-based learning environments, Web usage mining can help in analyzing and interpreting the learner's interaction with learning objects and resources on the Web in terms of basic activities recorded in log servers. For example, it is possible to analyze sequences of mouse clicks the learners made while browsing the Web pages related to a specific course (such as pages of lectures/presentations, pages containing links to lab materials, or pages related to examples of exam problems). The analysis can reveal what specific links the learners have followed, how long time they spend on each page, and what were entry and exit points to/from the page. Further interpretation of such statistics can lead to discovery of patterns like the following [36]:

- 84% of learners followed the expected pattern for tutorial elements;
- reduction in scaffold usage to 35% over time (scaffolding expects learners to become self-reliant).

Discovery of such patterns may result in understanding the learner better and converting this understanding into improved instructional design. The good thing is that Web usage mining enables observation-oriented interpretation and evaluation of the learners' behavior – it is objective, non-interruptive, always available, and can be automated. In practice, however, it should be complemented with other techniques as well.

At the moment, Web-based CSCL systems do not exploit Web mining potentials in their full capacity. Still, there are a number of Web mining challenges for CSCL systems on the Web. They define several CSCL-specific Web mining tasks:

- comparison – comparing expected and actual behavior of peer learners, expected (supported) activity and time spent on a page, and ordering/sequencing of actual basic activities to expected ones;
- classification – identifying the most likely purpose(s) of a learning session by analyzing the learner's actions;
- time series analysis – detecting changes in learner's behavior, learner profiling by means of detecting his/her learning strategy (the strategy is often based on changes in activities!);
- profiling learning time dimension – for example, discovering patterns like "60% of peers communicate intensively only during the first 10 minutes when solving problem A";
- discovering learning communities – this is possible if the Web is analyzed as as a huge directed graph in which a specific hub-authority-linkage pattern is a community's "signature"; essentially, this Web-mining task means scanning the Web graph systematically to locate such structures.

4.7 Evaluation Issues

No CSCL system can be widely accepted without a thorough evaluation. There are both general and CSCL-specific evaluation goals, problems, strategies, approaches and techniques that each CSCL evaluation study faces.

4.7.1 Evaluation Methodology for Learning Systems – General

There can be many different goals of learning system evaluation [9]; for example:

- to measure the performance of a learning system as a whole;
- to evaluate individual components of the system (e.g., the learner modeling component);
- to evaluate a certain learning strategy, theory or model underlying the system;
- to analyze the feedback to the student;
- to measure the system's navigation support for the learner;
- to measure adaptivity with respect to a selected learning goal and/or parameter;
- to obtain statistics of using parts of the domain knowledge.

In either case, there are some common problems of learning systems evaluation. Underlying theories are either new or still under development; there is no widespread agreement as to how the fundamental learning tasks should be performed. Also, there are no many truly operational learning systems in the same domain. Hence it is difficult to conduct an evaluation study based on comparison with other similar systems.

Typical approaches and techniques of evaluation include:

- conducting specifically designed learning experiments with the system and human learners, and having a human expert evaluate the results and the learners' interactions with the system;
- tracking the knowledge paths the learners take in the knowledge base;
- sensitivity analysis – values of some features of the system and/or system parameters are varied in order to determine which ones make the most difference in the system performance and learning effectiveness;
- simulated students – these are computer models (simulations) of human learners. The advantage of using simulated students is a higher level of control of the experimental setting, but it should be just a first step towards a full evaluation, since simulated students are too simplistic in comparison to real students in realistic environments.

Recent advances in learning technology and its integration with AI techniques has brought some new approaches as well – evaluation of educational services, evaluation of effects of enriching knowledge bases through machine learning, evaluation of educational ontologies, and ontology-based evaluation of the system's knowledge validity.

Generally applicable metrics used in evaluation studies of learning systems include total interaction time the learner spent with the system, the number of attempted problems, the number of solved problems, total number of attempts to solve the problems, problem solving speed [problem/min], accuracies of mastered and unmastered parts of domain knowledge encoded in the system, and many more. See [9] for a more elaborate discussion.

4.7.2 CSCL Specifics

Issues specific to CSCL systems are also evaluated by having a human expert judge how "reasonable" is the advice given by the system, in this case to the group of learners. This can be done by analyzing events in which several candidate advice with different rankings exist, and comparing them to human expert rankings using a suitable Euclidean measure of distance between the system's and the expert's rankings.

Specific evaluation metrics in collaborative learning scenarios include voting tracking, waiting-for-feedback timeout, participation balance for the group members (e.g., the number of contributions to the problem solution made by each member and the quality of contributions themselves), chat tracking, the time the group took to develop the case study solution, the difficulties in the prerequisite knowledge the group members should have acquired beforehand, the number of misunderstandings concerning the case study, matching of the group members' learner models, and discussion encouragement intensity. Note that researchers in the CSCL community stress that the ability to use and match the models of multiple learners connected to a collaborative learning systems is the key of adaptive collaboration support.

Recent research efforts have brought some new approaches to Web-based CSCL systems evaluation:

- Comparing students' individual and group solutions [5]. Conflicts between individual and group solutions constitute learning opportunities, provided that students recognize and address these conflicts. CSCL systems can be evaluated in terms of how successfully they encourage such negotiation when differences are detected, and how successfully they encourage participation in other ways.
- Evaluation of the participants' collaborative activity on the basis of their interactions [18]. This comprises mining logged data related to collaboration activities and using an intelligent technology (such as an expert system or fuzzy logic) to evaluate the collaboration. A necessary prerequisite for this approach is to design the CSCL system in such a way that the learners enter data used for evaluation along with problem-solving input. The drawback is that such a design can make the user interface unnecessarily more complicated.
- Web Mining-based evaluation (log data analysis) [7]. This kind of analysis can show the preferred and/or rarely kinds of online interactions among the peers during the learning sessions. For example, it may turn up through log data analysis that online interactions like discussion board postings and file uploads are infrequent compared to interactions through email messages; if the goal is to enforce the infrequent interactions, the system design can be improved in that direction. Also, analysis of log data related to individual contributions may

reveal group leaders, lurkers, and possibly (un)balanced contribution of different participants.

- Evaluation-supportive interfaces – making the learners enter data for evaluation automatically.
- Evaluation on both cognitive and metacognitive levels.

4.8 Summary

Current intelligent Web-based CSCL systems integrate a number of Internet and AI technologies. This is not to say that learning theories and instructional design issues should be given lower priority than technological support; on the contrary, new technology offers more suitable ways for implementing and evaluating instructional expertise in CSCL systems.

However, it becomes more and more obvious that the emerging Semantic Web makes the evolving infrastructure for future CSCL applications on the Web. Moreover, intelligent Web services and Web mining open a number of new possibilities to make Web-based collaborative learning more effective and more learner-centered. It is necessary for developers of Web-based CSCL systems to recognize such facts if they want their systems to be competitive enough and become widely accepted by the learners. Technology *does* matter.

Web Resources

In addition to published material on intelligent Web-based CSCL, there is a number of useful Web resources (usually maintained by dedicated research groups) that can be used as starting points for further research in theory and practice of intelligent Web-based CSCL. Some of them are:

COLLIDE Group (COLlaborative Learning in Intelligent Distributed Environments)
- http://www.collide.info/

LILT Group (Laboratory for Interactive Learning Technologies)
- http://lilt.ics.hawaii.edu/lilt/

ARIES Group (Advanced Research in Intelligent Education Systems)
- http://www.cs.usask.ca/research/research_groups/aries/

Mizoguchi Lab
- http://www.ei.sanken.osaka-u.ac.jp/

LTSN generic center, U. of Shefield, UK
- http://www.shef.ac.uk/collaborate/collaborative_elearning/index.shtml

ISO/IEC JTC1 SC36 Organization WG2: Collaborative Technology
- http://collab-tech.jtc1sc36.org/

References

[1] Adler, R.M. (1995). Emerging Standards for Component Software. IEEE Computer, March 1995, pp. 68-76.

[2] Andriessen, J., Baker, M., and Suthers, D. (Eds.) (2003). Arguing to Learn: Confronting Cognitions in Computer-Supported Collaborative Learning Environments. Kluwer book series on Computer Supported Collaborative Learning, Pierre Dillenbourg (Series Editor). Dordrecht: Kluwer.

[3] Buschmann, F., Meunier, R., Rohnert, H., Sommerlad, P., and Stal, M. (1996). A System of Patterns. Chichester: John Wiley & Sons.

[4] Chakrabarti, S., Dom, B., Gibson, D., Kleinberg, J., Kumar, S., Raghavan, P., Rajagopalan, S., and Tomkins, A. (1999). Mining the link structure of the world wide web. IEEE Computer, Vol.32, No.8, pp. 60-67.

[5] Constantino-González et al. (2002). Coaching Web-based Collaborative Learning based on Problem Solution Differences and Participation. International Journal of Artificial Intelligence in Education, IJAIED, Vol.13, pp. 37-63.

[6] Cooley, R., Mobasher, B., and Srivastava, J. (1997). Web Mining: Information and pattern discovery on the World Wide Web. In Proceedings of the 9th IEEE International Conference on Tools with Artificial Intelligence, ICTAI'97, pp. 321-330.

[7] Curtis, D.D., and Lawson, M.J. (2001). Exploring Collaborative Online Learning. Journal of Asynchronous Learning Networks, Vol.5, No.1, pp. 21-34.

[8] Devedžic, V. (2003). Key Issues in Next-Generation Web-Based Education. IEEE Transactions on Systems, Man, and Cybernetics, Part C – Applications and Reviews, Vol.33, No.3, pp. 339-349.

[9] Devedžic, V. (2003). Think ahead: evaluation and standardisation issues for e-learning applications. International Journal of Continuing Engineering Education and Lifelong Learning, Vol. 13, Nos. 5/6, pp. 556-566.

[10] Devedžic, V., and Harrer, A. (2004). Common Patterns in ITS Architectures. Accepted for publication in Kunstliche Intelligenz, Special Issue on Software Engineering for Knowledge-based Systems, No.3/04, pp. 17-21.

[11] Devedžic, V., Radovic, D., and Jerinic, Lj. (2000). Innovative Modeling Techniques on Intelligent Tutoring Systems. Book chapter in Jain, L.C. (Ed.), Innovative Teaching and Learning: Knowledge-Based Paradigms, Physica-Verlag (A Springer-Verlag Company), New York, pp. 189-234.

[12] Dillenbourg, P., Baker, M., Blaye, A., and O'Malley, C. (1994). The evolution of research on collaborative learning. In P. Reimann & H. Spada (Eds.) Learning in Humans and Machines: Towards an Interdisciplinary Learning Science, pp. 189-211. Oxford: Pergamon.

[13] Dillenbourg, P., and Schneider, D. (1995). Collaborative learning and the internet. [Online]. Available: http://tecfa.unige.ch/tecfa/research/CMC/colla/iccai95_1.html (Current: June 2004).

[14] Dillenbourg, P. (1999). What do you mean by "Collaborative Learning"? In P. Dillenbourg (Ed.), Collaborative Learning: Cognitive and Computational Approaches, pp. 234-243. Amsterdam: Elsevier.

[15] Gamma, E., Helm, R., Johnson, R., and Vlissides, J. (1995). Design Patterns: Elements of Reusable Object-Oriented Software. Reading: Addison-Wesley.

[16] Gassner, K., Jansen, M., Harrer, A., Herrmann, K., & Hoppe, H. U. (2003). Analysis methods for collaborative models and Activities, In B. Wasson, S. Ludvigsen, U. Hoppe (Eds.), Proceedings of the CSCL 2003, pp. 411-420. Dordrecht: Kluwer Academic.

[17] Gaševic, D., and Devedžic, V (2004). Teaching Petri nets using P3. Accepted for publication in the Educational Technology and Society (2004). Forthcoming.

[18] Hadjileontiadou, S. J., Nikolaidou, G. N., Hadjileontiadis, L. J., & Balafoutas, G. N. (2004). On enhancing on-line collaboration using fuzzy logic modeling. Educational Technology & Society, 7 (2), 68-81.

[19] Hopkins, J. (2000). Component Primer. Communications of the ACM, Vol.43, No.10, pp. 27-30.

[20] Hoppe, U., Pinkwart, N., Lingnau, A., Hofmann, D., & Kuhn, M. (2002). Designing and Supporting Collaborative Modelling Activities in the Classroom. In: Information and Communication Technologies in Education Volume I, A. Dimitracopoulou (ed.), Proceedings of 3rd HICTE, 26-29/9/2002, University of the Aegean, Rhodes, Greece, KASTANIOTIS Editions, Inter@ctive, pp. 185-190.

[21] Johnson, W.L., Rickel, J., and Lester, J.C. (2000). Animated Pedagogical Agents: Face-to-Face Interaction in Interactive Learning Environments. International Journal of Artificial Intelligence in Education, Vol.11, pp. 47-78.

[22] Muehlenbrock, M., and Hoppe, U. (2001). A collaboration monitor for shared workspaces. In J. D. Moore, C. L. Redfield, & W. L. Johnson (editors). Proceedings of the International Conference on Artificial Intelligence in Education AIED-2001, San Antonio, TX, May, pp. 154-165. Amsterdam: IOS Press.

[23] Pinkwart, N. (2003). A Plug-In Architecture for Graph Based Collaborative Modeling Systems. In U. Hoppe, F. Verdejo & J. Kay (eds.): Shaping the Future of Learning through Intelligent Technologies. Proceedings of the 11th Conference on Artificial Intelligence in Education, pp. 535-536.

[24] Preece, A., and Decker, S. (2002). Intelligent Web Services. IEEE Intelligent Systems, Vol.17, No.1, pp. 15-17.

[25] Roschelle, J. and Teasley, S. (1995). The construction of shared knowledge in collaborative problem solving. In O'Malley, C.E., (ed.), Computer Supported Collaborative Learning pp. 69-97.

[26] Schmidt, D., Fayad, M., and Johnson, R.E. (1996). Software Patterns. Communications of The ACM, Vol.39, No.10, pp. 37-39.

[27] Soller, A.L. (2001). Supporting Social Interaction in an Intelligent Collaborative Learning System. International Journal of Artificial Intelligence in Education, Vol.12, pp. 54-77.

[28] Supnithi, T., Inaba, A. Ikeda, M. Toyoda, J., and Mizoguchi, R. (1999). Learning Goal Ontology Supported by Learning Theories for Opportunistic Group Formation. Proceedings of the International Conference on Artificial Intelligence in Education AIED-1999, Le Mans, France, pp. 263-272.

[29] Suthers, D.D. (2003). Representational Guidance for Collaborative Learning. (Keynote address for 11th International Conference on Artificial Intelligence in Education: AI-ED 2003). In: H. U. Hoppe, F. Verdejo, and J. Kay (Eds.) Amsterdam: IOS Press, pp. 3-10.

[30] Szyperski, C. (1998). Component Software – Beyond Object-Oriented Programming. Reading, MA: Addison Wesley Longman.

[31] Šimic, G., and Devedžic, V. (2003). Building an intelligent system using modern Internet technologies. Expert Systems With Applications, Vol.25, No.3, pp. 231-246.

[32] Vassileva, Greer, J., McCalla, G., Deters, R., Zapata, D., Mudgal, C., & Grant, S (1999). A Multi-Agent Approach to the Design of Peer-Help Environments. In Proceedings of AIED'99, Le Mans, France, pp. 38-45.

[33] Vinoski, S. (2002). Web Services Interaction Models, Part 1: Current Practice. IEEE Internet Computing, May-June, pp. 90-92.

[34] Vygotsky, L. (1978). Mind in Society: The Development of Higher Psychological Processes. Harvard University Press, Cambridge, MA.

[35] Wasson, B. (1998). Identifying Coordination Agents for Collaborative Telelearning. International Journal of Artificial Intelligence in Education, Vol.9, 275-299.

[36] Xu, L., Pahl, C., and Donnellan, D. (2003). An Evaluation Technique for Content Interaction in Web-based Teaching and Learning Environments. In Devedžic, V., Spector, J.M., Sampson, D.G., Kinshuk (Eds.), ICALT 2003 - Proceedings of The 3rd IEEE International Conference on Advanced Learning Technologies, Athens, Greece, July 09-11, 2003, pp. 347-352. Los Alamitos, CA: IEEE Computer Society.

5. Using Multiagent Intelligence to Support Synchronous and Asynchronous Learning

Xuesong Zhang, Leen-Kiat Soh, Hong Jiang, and Xuli Liu

Department of Computer Science and Engineering University of Nebraska, Lincoln, NE, E-mail: {xuzhang, lksoh, jiang, xuliu}@cse.unl.edu

This chapter presents an innovative multiagent system to support synchronous and asynchronous cooperative learning both in the real classrooms and in distance education. The system, called I-MINDS, consists of a group of intelligent agents. A teacher agent monitors the student activities and helps the teacher manage and better adapt to the class. A student agent, on the other hand, interacts with the teacher agent and other student agents to support cooperative learning activities behind the scene for a student. Two I-MINDS innovations are (a) agent-federated real-time "buddy group" formation and refinement, and (b) automated ranking of questions and responses. These two functionalities are supported by a suite of knowledge bases, applied to data that are either collected and derived from the agent-mediated activities, or compiled directly from online student surveys. The knowledge bases include instructional keywords and rules for profiling student and scoring questions. We have tested our I-MINDS prototype within a pilot study. In this pilot study, we had two groups: control and I-MINDS. Each group was given two lectures by the same instructor on GIS topics. The result was very promising as were comments from the instructor and the subjects in the I-MINDS group related to their comfort level in using the tool.

5.1 Introduction

Information technology is rapidly changing the educational process by enhancing the way information and knowledge are represented and delivered to students. The advent of Internet and multimedia technology has meant potentially drastic changes in the teaching and learning process from the traditional classroom setting to a more geographically distributed, virtual but still interactive one.

In this chapter, we describe an innovative multiagent system that supports cooperative activities among students through the use of intelligent agents. This system, applicable to students both in the classroom and distance education, monitors students' and teacher's activities to help the teacher teach better and students learn better.

As pointed out in [13], research strongly supports the use of technology as a catalyst for improving the teaching and learning environment. Educational

X. Zhang et al.: *Using Multiagent Intelligence to Support Synchronous and Asynchronous Learning*, StudFuzz **178**, 111–140 (2005)
www.springerlink.com

technology has been shown to stimulate more interactive teaching, effective grouping of students, and cooperative learning.

According to Sheppard *et al.* [24], the motivations behind using technologies in learning are related to the beliefs that they

- are inherently good
- are needed to remain competitive as an institution
- make the delivery of education more cost effective
- open up possibilities of reaching new/different student groups
- offer students more control over when and where they interact with "knowledge"
- offer more and new opportunities for student-student and student-faculty interaction
- offer new opportunities for cross-university interaction by both students and faculty
- offer students richer, more diverse learning resources and alternative points-of-view
- offer opportunities for documenting, cataloguing and re-using curriculum materials and student work

Our system, called the Intelligent Multiagent Infrastructure for Distributed Systems in Education (I-MINDS), incorporated technology to support and encourage cooperative learning among students in a real-time classroom, for either distance education or in-class students. The motivations behind our project are three-fold: (a) to address or improve areas in education (e.g., distance education) in line of the reasons (competitiveness, cost-effectiveness, student control, student-student interactions, student-faculty interactions, and management of curriculum materials and student work) outlined by Sheppard [24] above, (b) to investigate and ultimately adopt innovative techniques in Computer Science (such as multiagent intelligence) and Computer Engineering (such as distributed computing) in the advancement of education tools, and (c) to investigate cooperative learning from the viewpoint of instructional technology and action research, by using I-MINDS as an active, flexible testbed.

I-MINDS design, at the cognitive level, exploits the intelligent characteristics (i.e., reactivity, pro-activeness, and social ability [32]) of agents and multiagent systems to better support a real-time classroom environment. Our teacher and student agents collaborate to better serve the teacher and the students. A teacher agent helps a teacher better manage the real-time classroom and analyze post-lecture statistics—ranking questions, delivering lectures, profiling students, etc. A student agent interacts with a student to address his or her needs. It also profiles the behavior of the student and other student agents, making use of these profiles to form effective "buddy" groups behind-the-scene. It also interacts with the teacher agent to obtain data and information. Meanwhile, the student agent, working together with the teacher agent, archives and customizes the course content of every synchronous session to facilitate the student's asynchronous study. This design is different from most current agent-based education systems in which the full capability of agent intelligence is not used. Furthermore, by allowing the student agents to interact closely, we make use of multiagent system intelligence—each agent is not an isolated entity interacting with only the teacher or the student.

In the following, we first define what agents are and some related work in educational applications. In Section 5.3, we describe the design and methodology behind our teacher and student agents. In Section 5.4, we discuss the implementation of I-MINDS, including how two innovations are embedded into these agents to (a) automatically rank students' questions and responses, and (b) automatically form and refine "buddy groups" for the students. A buddy group, in our context, is a close-knit student group where its members exchange messages and help each other understand the lectures. Also, in this section, we will discuss the features for synchronous and asynchronous learning that make I-MINDS more than just a lecture-delivery tool. In Section 5.5, we report on the pilot study that we have conducted using our I-MINDS prototype, and show that the preliminary results are encouraging. Finally, we conclude this chapter with remarks on the main contributions and future research directions of our work.

5.2 Related Work

First, let us define some terms within the context of this chapter.

An *agent* is a module that observes and receives input stimuli from its environment, makes autonomous decisions based on these stimuli, and actuates actions to carry out these decisions, which, in turn, changes the environment [31].

An *intelligent agent* is a flexible agent that is pro-active, reactive, or social [32] and able to *learn* to improve its own performance [23].

A *multiagent system* is a group of agents that interact with each other to accomplish tasks [31]. When we talk about *multiagent intelligence*, we refer to the capability of these agents learning from each other and learning from the system environment to better accomplish tasks. The benefits of using multiagent intelligence include the sharing of experience and knowledge, and speeding up of machine learning. For example, suppose that agent *A1*'s responsibility is to support students on how to use an online browser. From its database, agent *A1* knows how to explain the features of browser *B1*. Now, there is a student who wants to learn about browser *B2*. Agent *A1* may learn about the features of browser *B2* from investigating the browser itself and its knowledge of *B1*, or it may communicate with other agents in the system to see whether any agent has an explanation already stored in its database, and if so, to share that information. This allows the multiagent system as a whole to be more responsive to tasks and allows each individual agent to learn from others.

There are in general two approaches to apply agent-based technology in education systems: (1) as an individual intelligent agent, such as intelligent tutoring systems, and (2) as a group of intelligent agents in a multiagent system environment.

Most software tutors (or Intelligent Tutoring Systems) [8] are based on knowledge-based systems or single agents. For example, SAM [5], intelligent interfaces such as AIP, PIP, and IMP [3], PAT [22], Animated Pedagogical Agents [12], AutoTutor [9, 13, 16], PACT [2, 14], ANDES [7], SHERLOCK [15], Herman the Bug, and Cosmo the Internet Advisor [16] utilize single agents to observe and adapt to the different behaviors of students. Because these agents work independently of each other, they fail to utilize the potential of multiagent intelligence. The advancement of multiagent systems presents a major opportunity to develop an infrastructure in which agents communicate, exchange experiences, and cooperate to better serve the instructors and students. I-MINDS exploits this opportunity. I-MINDS design focuses on the agents serving the teachers and students, profiling their behaviors to more effectively take advantage of the class time. For example, the teacher agent ranks questions that are likely to be useful for the teacher, and the teacher is able to pick the best ones to answer in class. Similarly, all students are able to post questions as often as desired. This creates a very rich interaction that is logistically impossible in traditional classrooms. In addition, a student agent is capable of forming and refining the "buddy" group for its corresponding student behind-the-scenes. Therefore, students are able to interact actively without disrupting the classroom and in a manner less constrained by space and time.

Only a few systems make use of multiagent intelligence, and those that do only monitor different agents to conduct team performance analysis, such as PROBES [20], ISSAC [22], and the automated assistants of [21] to help humans understand team behaviors. I-MINDS agents move beyond monitoring to actively and autonomously seek out buddies for their users and learn to refine the buddy group.

Finally, one of the abilities featured in intelligent agents is self-improvement [33]. That is, agents should be able to *learn* to help teachers teach better and to help students learn better—some have even argued that only agents that learn are truly intelligent agents [32]. I-MINDS agents learn: the teacher agent learns how to evaluate the students and how to evaluate the responses of the students. The student agents learn how to form more effective buddy groups and how to help their fellow students in active and cooperative learning processes.

Some synchronous virtual classrooms are currently being developed by research institutions and companies, such as IRI-h [1], Centra Symposium [6], Interwise E-Learning [10], and Mimio Classroom [30]. IRI-h is designed to function within heterogeneous network environments, and offers audio, video and tool sharing services. It can also support class participants with limited multicast capabilities, or limited connectivity bandwidth by providing a scalable infrastructure. As a commercial product, Centra Symposium has many practical features, such as structured live interaction, asynchronous learning, rich content support, low bandwidth requirement, enterprise-class management and scalability, ease of deployment. Interwise E-Learning can offer one-on-one mentoring sessions, deliver live classes, hold collaborative learning sessions, and populate a knowledge repository with on-demand learning objects. Mimio Classroom allows the students to share notes with the teacher in real time, and students are able to add their own individual comments and notes, which can be saved and reviewed later. I-MINDS has most of the above features, and exploits multiagent intelligence to better utilize the information. This provides I-MINDS with the unique capability of utilizing the multiagent technology to add intelligence to the system to effectively help the teaching and learning process.

5.3 Design

In this section, we describe the design of I-MINDS. Our design focus is two-tiered. The system-level objective is to build an infrastructure that is scalable, flexible, and portable. The agent-level objective is to build an infrastructure that allows agents to be reactive, pro-active, and social to ultimately collaborate effectively. For details of our distributed design, please refer to [18, 19].

Figure 1 depicts the logical infrastructure of I-MINDS. I-MINDS infrastructure is based on a layered model where each layer provides a distinct set of functionality for the system. The separation in this fashion provides a well-organized model that facilitates the development and future enhancements. The network underlying the fundamental communication environment serves as the first layer. System-level protocols and encapsulations equip the second layer to provide convenient communication and deployment functions to upper levels. To facilitate the communication programming, OO programming model (e.g., Java socket programming) and DSO [17] are provided as the third layer. Finally, the agents are located at the uppermost two layers. Each agent has two interacting modules: content-independent and content-dependent. The content-independent module provides the definitions and processes for education-related, general services, while the content-dependent module handles specific course-related knowledge base, heuristics, and data. This design allows our system to be highly flexible and user-friendly. Also, since I-MINDS is developed using Java, it has high portability and is able to work on heterogeneous environments.

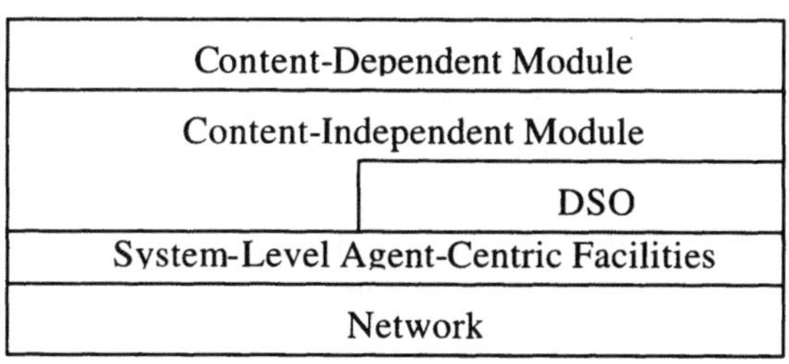

Fig. 1. The Logical Infrastructure of I-MINDS

Figure 2 depicts the topological structure of I-MINDS. Considering that some remote students (and thus their respective student agents) may access the Internet using slow dial-up connections, and multi-cast may be prohibited in some network configurations, we designed the topological structure of I-MINDS as shown in Figure 2. On the top is the manager of the system, which has the overall information of the system, such as all the courses provided, currently ongoing classes, and other static and dynamic information about student registration. The teachers (and their corresponding teacher agents) are located at the second layer, and they give the lectures. Besides the students (and their corresponding student agents), there are some remote proxy servers at the third layer, which serve as bridges between the teachers and the students from the network where there is low-speed access only or multicast is prohibited. The function of the remote proxy server is similar to the Gateways [1]. Students access the virtual classroom through the nearest remote proxy server indirectly (e.g., student 4), as denoted by the dotted, curvy lines.

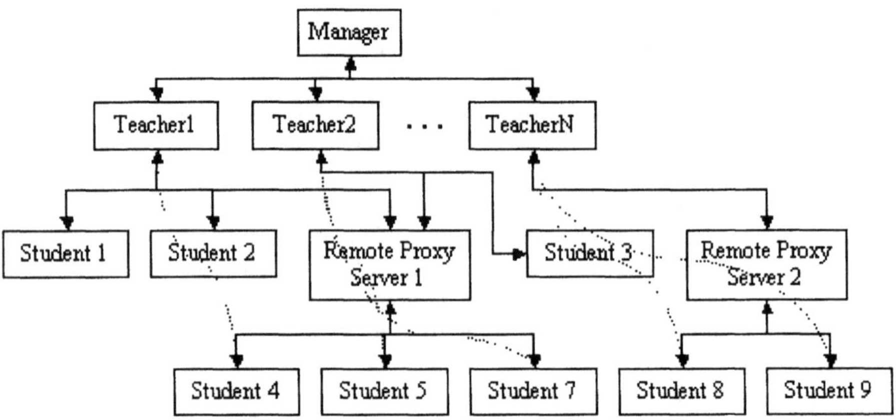

Fig. 2. The Topological Infrastructure of I-MINDS

5.3.1 Design Features

I-MINDS is designed to achieve the following features:

- Portability. We implemented I-IMINDS using Java, so it would execute on any operating system that supports a JVM.
- Scalability. By using the remote proxy servers shown in Figure 2, we can not only easily expand I-MINDS' scalability, but also support those who have limited Internet connectivity or are located in networks without multicast capabilities.
- Flexibility. Each module in the layered infrastructure shown in Figure 1 can be expanded or adjusted independently without affecting the functionality on other layers. Moreover, this infrastructure supports different course subjects by transplanting the content-dependent modulus.
- Intelligence. Multiagent system intelligence is employed in I-MINDS, each attendee in the virtual classroom is equipped with an intelligent agent to help

the teaching and learning process, and moreover, these agents also collaborate with each other.

- Low bandwidth requirements. Instead of making the teacher's server take care of everything, with the help of an intelligent agent, the student's computer can do a lot of work, and thus the communication overhead will be reduced effectively. We further reduce the communication overhead by designing an algorithm to trace handwriting of the teacher on the whiteboard, and transmit only the trace of the teacher's pen to the students, rather than capturing a video stream like other on-line education systems. According to our experiment, the bandwidth requirement is only 17 kbps with the audio and teacher's handwriting on the whiteboard.
- Low hardware requirements. The necessary devices for the teacher only include a high-performance computer and a microphone. To achieve better teaching effect, the teacher may also choose a Mimio, a Mimio mouse, a LCD projector, and a Webcam. As for the student, a personal computer with the standard equipment such as microphone and speakers will provide all the necessary functionalities and equipment for full access to I-MINDS resources.

5.3.2 Multiagent Systems

The development and implementation of I-MINDS is a unique and innovative approach to computer-aided instruction and learning because of the incorporation of active and intelligent software agents. I-MINDS includes both student agents and teacher agents. Student agents collaborate to autonomously form student *buddy groups* and provide intelligent services to their respective student users. Teacher agents monitor classroom activities and analyze student behavior to help the teacher respond to questions and to assess student comprehension and interest. These intelligent agents are designed to assess their own performance based on the observed impact of the buddy groups and the agent-initiated interventions, such as question ranking, on student learning. For further details on the use of information among the agents to support a classroom, please refer to [28, 29].

The first I-MINDS innovation is in agent-federated "buddy group" formation. A buddy group *is a team* (or "coalition") that is formed dynamically to support the members, or buddies, within the group to achieve common goals. Student agents, profiling the activities of their respective students/learners, seek out compatible student agents to form buddy groups, allowing a free-flow of questions and answers between members of the same buddy group. A "good buddy" with good responses will be ranked high and preferred for close collaboration. A "poor buddy" who never responds, for example, will be removed from the buddy group. We plan to incorporate a cooperative learning model [11] using proven concepts to extending the expertise and intelligent components of our I-MINDS agents in the future.

The second innovation is the automated ranking of questions and responses for the instructor using agent intelligence. Each question or response from a student will be analyzed and ranked for the instructor according to its appropriateness, quality of information content, etc. Currently, such rankings are based on keyword selection. This selection also teaches the agent to learn to evaluate questions better, making the agent highly adaptive and dynamic. A key component of this innovation is that it encourages interaction and, thus, active learning [4, 26]. As noted in Section 5 on the proof-of-concept study, questions asked through I-MINDS tended to have higher quality and required more information-rich responses than those asked by a control group in a traditional classroom setting.

5.3.3 Teacher Agent

Figure 3 shows the structure of the I-MINDS teacher agent.

In the Content-Dependent Module are quizzes/exercises and answers from all the students, questions asked by students, rules used for inference, and dynamic profiles of the students. The profile of a student is a score based on the student's questions and cooperative learning activities. The I-MINDS teacher agent automatically monitors and logs these into its database. The profiles allow the teacher agent to rank questions and help the student agents build buddy groups. The rules used to evaluate the quality of the questions and the buddy groups will be evolved by I-MINDS based on their utility. The Evaluation mechanism evaluates the students based on their responses to the exercises and the quizzes as well as the monitored questions and actions from their student agents. These profiles also factor into the Self-Learning activity. For example, the teacher agent will be able to learn which keywords and heuristics are useful to improve the quality of questions from the students. Finally, a Repository Management mechanism caches sizable teaching materials into large storage devices for efficient transmission.

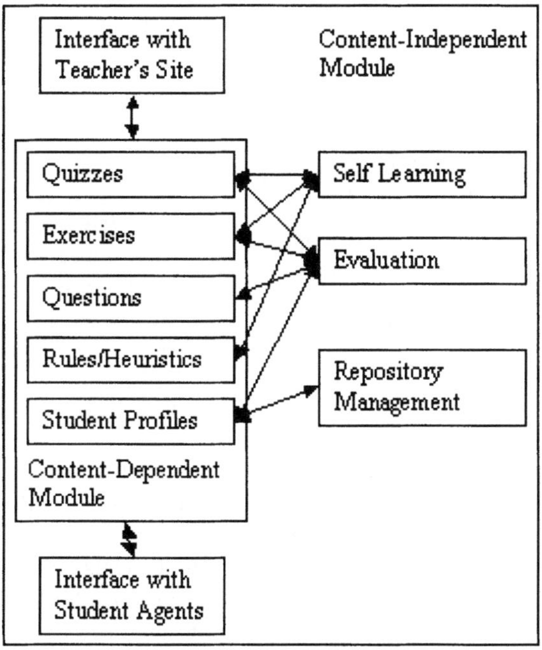

Fig. 3. The Structure of the Teacher Agent

Currently, we have implemented most of the modules such as questions, rules/heuristics, and student profiles of the teacher agent and have implemented both of the Interface modules and part of the Evaluation module. We have also implemented a prototype of the Student Profile module and a definition of a student profile. The response by each student through his/her respective student agent is evaluated, and the cooperative learning activities among students are captured automatically by the teacher agent.

5.3.4 Student Agent

Figure 4 depicts the structure of the I-MINDS student agent. The student agent displays messages and information streams received from a teacher agent directly to the student. Similarly, the student agent forwards the responses from the student to the teacher agent.

The Tracking mechanism tracks the activities and the progress of the student. For example, if the student does not touch the keyboard or move the mouse for five minutes during a class, the student agent may play a sound to alert the student to concentrate on class. If the student misses one class, the Tracking mechanism may go to the corresponding teacher agent, find the archived materials for that class according to the timestamps or the syllabus the teacher provides, and remind the student about the missed lectures.

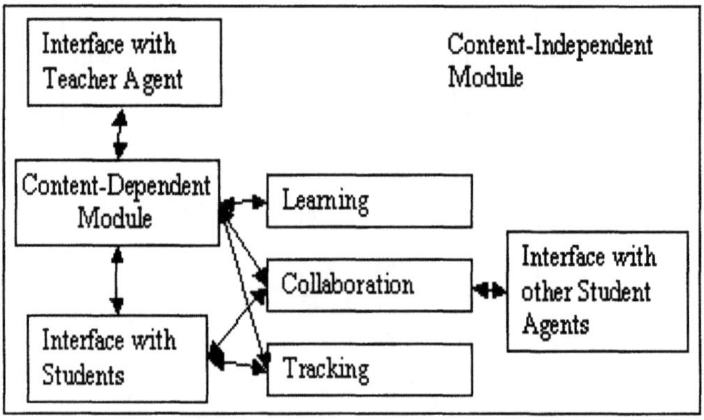

Fig. 4. The Structure of the Student Agent

Each student agent has a Collaboration mechanism that can be activated by the student. When a student asks a question, the student agent sends it to the teacher agent. In addition, the student agent sends the question to the other student agents identified as buddies by the first student's agent. Thus, buddies may answer questions that the teacher does not respond to in class. Buddies that have not been responsive will be dropped from the buddy group; buddies that have been helpful will be approached more frequently. The student agent performs these tasks for its user autonomously.

To date, we have completed most modules of the student agent, including the Collaboration module and the Tracking module. The I-MINDS student agents are able to form buddy groups dynamically based on the information shared among the student agents. Thus, the I-MINDS design employs multiagent intelligence to form and refine cooperative learning teams.

5.3.5 Intelligent Module

Our teacher agent and student agents are equipped with machine learning mechanisms. The teacher agent aims to learn to better profile the student agents that it interacts with, including the assessment of the student performance that the student agents represent, the buddy group activities, and the response of the students through the student agents. As will be discussed later in Section 5.4.1.4, the teacher agent supports the teaching process by automatically scoring and ranking the questions asked by the students through the student agents. Scoring and ranking these questions are based on several weighted factors. The teacher agent has machine learning mechanisms in place to refine these weights to more accurately evaluate the questions. From a student agent's point of view, it wants to learn to support the learning process, in particular the buddy group activities, as will be discussed later in Section 5.4.1.6. Presently, we have designed and implemented an end-to-end machine learning module (shown as the "Self Learning" module in Figure 3) for the teacher agent while that for the student agents is still under construction. Here we describe briefly how the machine learning mechanism works.

In the current Intelligent Module design, to score and rank a question, the agent uses three sets of heuristics: keyword-based, student profile-based, and question-based. When the teacher agent receives a question, it first parses the question into individual words or tokens. Then it matches the tokens against the keywords stored in the agent's knowledge base for a particular class, and a particular lecture. It also brings up the profile on the student who asked the question. The student's past behavior and quality of questions that he had asked contribute to the score of the current question. Finally, the question's format, length, and timing also play a role in its score. All the positive and negative contributions to a question's score are recorded. Now, when the question is presented to the teacher (see Section 5.4.1.4), the teacher may choose to discard the question, answer the question, or delay dealing with the question while choosing to answer other questions first. Whenever a question is chosen, the teacher agent realizes that the teacher has just provided a valuable feedback to the way the teacher agent has scored the question. Similarly, whenever a question is discarded, the teacher agent also processes that action or choice as a feedback to its reasoning process. Now, if a question is ranked low and scored low, and yet it is chosen out of order (over other questions of higher ranks and higher scores, then the teacher in essence signals to the teacher agent that the agent has scored the question too low. If a question is ranked high and scored high, and yet it is discarded, then the teacher agent takes it as a hint that it has scored the question too high. As a result, the teacher agent traces back to the positive and negative contributions of that question. If the question has been scored too high, then it lowers the weights of the heuristics that contributed positively to the score of the question and increases the weights of those that contributed negatively; and vice versa. This way, if the same question is encountered in the future, the agent will be able to score and rank better.

For the keyword-based heuristics, a teacher has to supply a set of keywords deemed important to the upcoming lecture. Each keyword is also associated with a weight. The teacher agent learns to refine these weights according to the teacher's actual decision during the real-time lecture and also adds new keywords to the list. Thus, in a way, the teacher agent *mines* and *discovers* keywords that are important but not included in the initial set of keywords through reinforcement learning, provided by the teacher real-time. As the lecture progresses, the teacher agent will refine the weights of the keywords (initial and new). When the lecture concludes, the teacher agent will have a set of new keywords. This machine learning mechanism thus lends true intelligence to the teacher agent—rendering the agent the power to improve its own performance.

For the student profile-based heuristics, a set of initial rules based on self-efficacy, motivation, and aptitude is stored with the agent. For example, one such heuristic is based on the observation that students with a high aptitude and motivation usually ask better questions during the lecture. Each heuristic is also tagged with a weight indicating the confidence value in the assertion made by that heuristic. Similarly, the teacher agent modifies the weights of the heuristics based on the feedback received from the teacher.

We also have the same treatment for the question-based heuristics. Long questions are considered, in general, better than short questions, and questions that come right after a lecture page are considered, in general, better than questions that come a while after a lecture page. Similarly, agent modifies the weights of the heuristics based on the feedback received from the teacher.

Our preliminary tests showed that this Intelligent Module of the teacher agent is able to learn new keywords, refine the weights of the heuristics, and record the trends on how the weights change for each heuristic.

Currently, we continue to improve the Intelligent Module to meta-learn its machine learning behavior and a set of buddy group-based activities.

5.4 Implementation

In this section we describe the system configuration for both synchronous learning and asynchronous learning in I-MINDS. In the synchronous section we talk about how teachers and students use I-MINDS in a way that is analogous to the traditional classroom, how intelligent agents help the teaching and learning process, and how the students cooperate with each other through the intelligent agents. In this section we also discuss how course materials and events, like questions and answers, of a class session are archived. In the asynchronous learning section, we describe how students review, manage, revise, and replenish locally archived course materials and use archived course materials to replay the class session.

5.4.1 Synchronous Learning in I-MINDS

In synchronous learning sessions, both teacher and students log in to the system at the same time and can communicate in real-time. I-MINDS aims to provide the same quality of experience for both teacher and students as they get in traditional classroom, plus real-time and behind-the-scene intelligent agents support.

5.4.1.1 System Environment

Figure 5 depicts a typical I-MINDS system setup. At the teacher's site, the teacher is equipped with an array of computing and multimedia components: (a) the I-MINDS teacher application, (b) any desktop/laptop computer with Pentium 1.7 GHz or faster CPU and 256 MB or more memory, (c) high speed Internet connection, such as T1 network in office or broad-band ISP at home, (d) a microphone to capture audio, (e) a webcam to capture video. The following components are optional: (f) a Mimio and a Mimio mouse (Virtual Ink Corporation) to capture the teacher's handwriting on a flat surface such as a whiteboard, (g) a whiteboard (or a flat surface), and (h) a LCD projector to project lecture slides onto the whiteboard for the teacher to annotate on and for on-site students to view.

The student's site requires only an average-performance computer to run I-MINDS student application, and optional microphone and webcam to carry out real-time video/audio communication with the teacher and fellow students. Contrary to the high speed Internet connection required on the teacher's site, dial-up speed is adequate for students to utilize the basic functionality of I-MINDS.

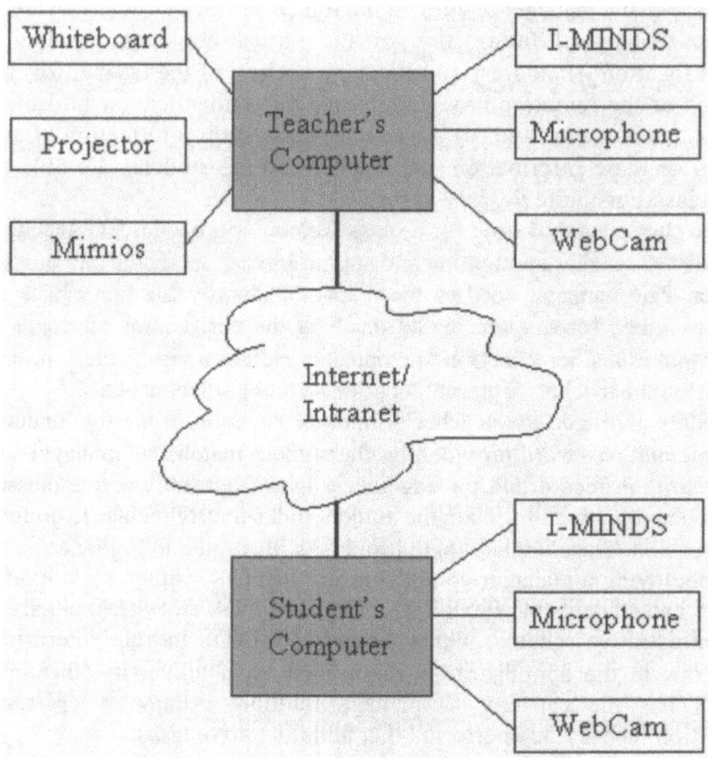

Fig. 5. The System Setup of I-MINDS

5.4.1.2 Connecting to the system

I-MINDS provides two ways to launch the student application. One is the stand-alone installation version, in which students get a copy of the installation program and install I-MINDS on to their computers. Then students are able to configure and run the I-MINDS student application just like any other applications installed on their computers. The other way is using the web start version, where students go to the I-MINDS webpage and launch student application by one click on the hot link. This version is designed to accommodate students who travel often, or do not have the privilege to install applications onto the computer system. Since this version does not require installation, it is usable through publicly available computers, like the ones in public libraries or a hotel's lobby room.

Unlike many other distance education systems that require special equipments and the teachers to go to a specific classroom to conduct a lecture, I-MINDS gives the same flexibility that students experience to teachers. With I-MINDS, teachers can use any decent computer with high speed Internet connection to give lectures from anywhere, home or office.

To connect the geographically and dynamically distributed teachers and students into one system, I-MINDS utilizes a manager component. From Figure 2 we can see that a manager resides on the top of this infrastructure, which holds the dynamic topological information of the virtual classroom, including (a) the Internet location (IP address and listening socket) of the teacher, (b) the Internet locations of the remote proxy servers, (c) the authentication information of all teachers and students, and (d) the course registration information of all students. With all of these information, the teacher and the students are able to join the virtual classroom quite flexibly as follows.

When the scheduled time for a class session approaches, the teacher will start the I-MINDS teacher application and submit his/her username and password to the manager. The manager verifies the teacher's identity and gives back a class list that is assigned for this teacher to teach, if the verification succeeds. After the teacher makes his/her selection, the manager creates a virtual classroom session by associating the teacher's Internet location with the selected class.

Students will get the teacher's Internet location from the manager. If the username and password provided by the student match, the manager will look up the registration record and put together a list of currently active classes that the student had registered in. Then the student picks the right class to connect into the active session. The complete login process is illustrated in Figure 6.

By deploying a manager component in I-MINDS system, we not only give the teachers and students the flexibility to use I-MINDS anywhere, but also introduce an administrative scheme, which elevates I-MINDS into an enterprise solution. Analogous to the administration department of a university, the manager of I-MINDS has the capacity to manage multiple groups of classes, maintain registration records, and perform other administrative tasks.

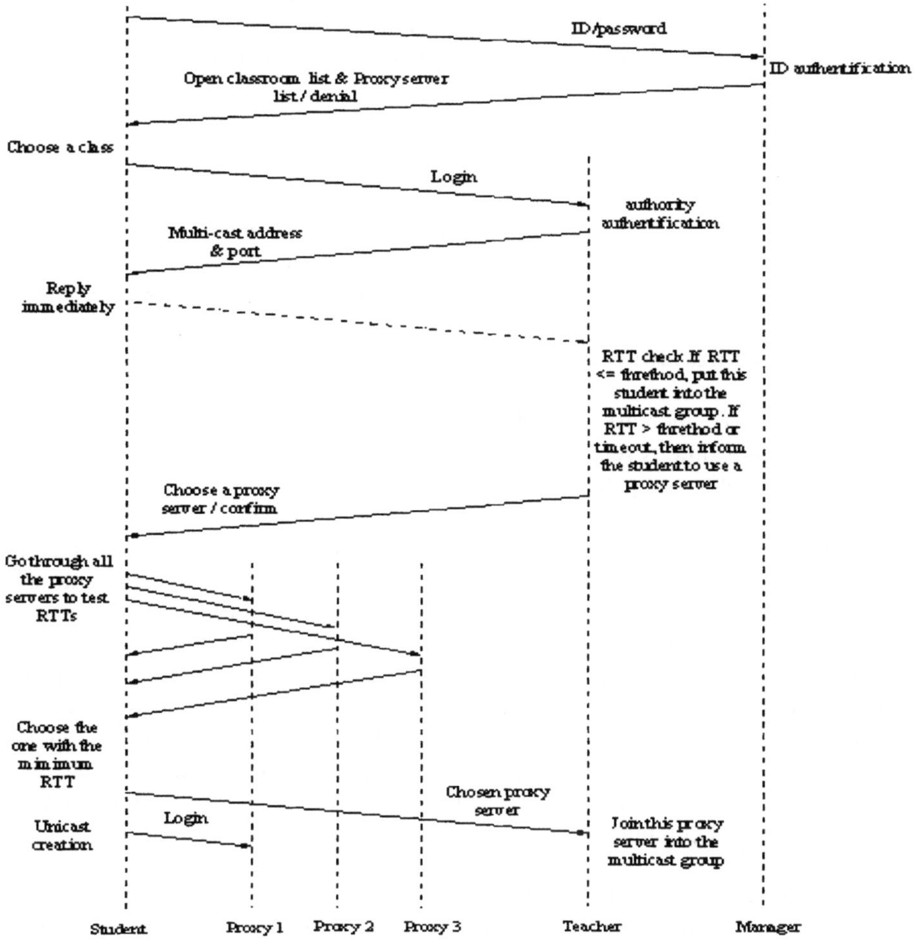

Fig. 6. The Login Process

5.4.1.3 The Teaching and Teacher Agent Processes

According to [25], many professors in traditional universities are reluctant to teach distance education classes. One prominent reason is that in most distance education systems, the instructors are required to prepare the course materials in a particular format. Some programming skills, as well as art design, are often necessary to develop good course materials. I-MINDS solved this problem in an innovative way.

In I-MINDS teachers are not required to prepare special course materials. The same material used in teaching traditional classes can be deployed directly in I-MINDS without any modification. Further more, I-MINDS allows teachers to use anything that displays on their computer screen as course materials. These additional materials may be a Microsoft Word document, a picture, a web page, the interface of a software application, or even a white screen to draw diagrams on. All of these can be done through a simple interface shown in Figure 7.

Fig. 7. The Interface of the Teacher's Application

The complete teacher's application comprises seven functional buttons. This interface stays on top of all other application windows on the teacher's computer screen, and thus it is always directly accessible.

NextSlide. To increase I-MINDS' scalability, as well as decrease the delay between when a slide is sent out from the teacher and when it is finally displayed at every student's computer, the teacher can compose a series of slides and distribute them to students prior to the class session. Then in class the teacher can use "NextSlide" button to turn to the next slide. At the same time the teacher agent will signal the student agents, which automatically synchronize students onto the same page. In this way, only an 8-byte integer is transferred instead of an entire slide that weights tens of Kbytes or more (in case of high-resolution pictures).

SendPage. When the teacher needs to add any extra material in between these prepared slides, he/she can first bring the material to be displayed on the computer screen, and then hit the "SendPage" button. Whatever is shown on the computer screen will be captured, saved into the format of a slide, encapsulated and transmitted to student agents. On student site, student agent will display the new slide on student's computer screen and save it into the prepared slide series in the appropriate order.

ModifyPage. The teacher is also able to annotate directly on top of a slide or even modify the content of a slide by using the *pen*, *eraser*, and *marker* tools, which are invoked by clicking the "ModifyPage" button. Any handwriting, eraser movement, and marker highlight are captured as the teacher clicks and drags the mouse, and transmitted, as a new layer of the current slide, to students when the teacher releases the mouse. The new layer is represented by an array of integers, which includes information like the tool's type, the tool's color, and the trace of the mouse movement. When the student agent receives such an array, it reconstructs the layer, resizes the layer according to the slide size displayed on student's computer screen, and merges the layer down to the slide to show the changes the teacher had made. Figure 8, captured in our pilot test, shows the instructor drew a circle (in red) on every bullet on the left side of his PowerPoint slide to accentuate students' attention as he went through the list.

BlankPage. By the same token, the teacher can bring up a blank slide with white background by clicking the "BlankPage" button, and use it as a scratch paper to draw diagrams or illustrate ideas.

It is often found awkward to draw or annotate with a mouse, or even a graphic tablet. Especially when the class also accommodates on-site students, it is impossible to always sit in front of the computer. One solution for this problem is to let the teacher write or draw on a traditional whiteboard and digitalize the marker's movement. I-MINDS is designed to be compatible with wireless input devices. In particular we use Mimio in I-MINDS' tests.

Mimio is an input device that consists of a sense-and-capture bar, installed on a flat surface, and a stylus that, when pressed against a surface (a whiteboard, for example), emits a signal at the frequency detected by the sense-and-capture bar. The signal is used to determine the stylus' location. In this way, everything written by the teacher on the whiteboard is captured and digitized automatically and fed into the computer connected to the Mimio.

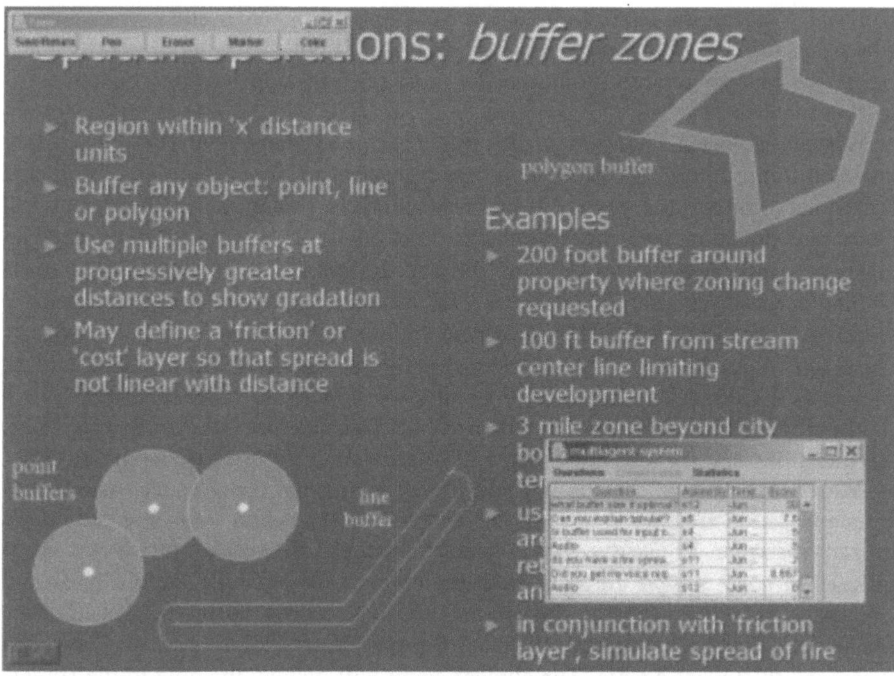

Fig. 8. A Snapshot of the Teacher's Computer Screen during Our Pilot Test

By projecting the teacher's computer screen onto a whiteboard and calibrating the projection through the Mimio, we convert the whiteboard into a large touch screen. The teacher can control the computer directly by touching the whiteboard and write on the white board using Mimio stylus, which serves as a wireless mouse. This multimedia setup not only enables I-MINDS to accommodate on-site and remote students at the same time, but also makes teaching a distance class as convenient and comfortable as teaching a traditional class.

Comparing to teaching a traditional class, teaching a distance class using I-MINDS is more convenient and efficient. No special course material is required to prepare, which means the same course material the teacher had prepared for traditional class can be directly used in I-MINDS to teach distance class. The teacher can also use anything on the computer screen as course material, and annotate or modify them. And all of these handy features can be accessed on top of a traditional whiteboard using the Mimio stylus.

5.4.1.4 The Support of Teacher Agent

In this section we describe how the teacher agent utilizes its intelligence besides delivery of course materials to students.

QuestionList. One of I-MINDS innovations is using software agents to automatically rank questions and responses. As pointed out in [25] distance education is time consuming in terms of answering students' questions, because students can ask unlimited number of questions any time, in class or after class. Especially during the lecture, the teacher cannot address each and every question when the number of questions is large. With the help of the intelligent teacher agent, the teacher is able to choose the most relevant and valuable questions to answer in class.

When a question is raised to the teacher, the teacher agent first passes this question to the Intelligent Module, where the question is analyzed, classified, and scored according to the keywords and heuristic rules. Then the teacher agent will insert the question into the sorted question queue according to its score. The teacher is informed of the presence of new question but not interrupted from his/her current process. The teacher can view the question queue anytime by clicking "QuestionList" button shown in Figure 7. Questions can be answered, discarded, or marked as a repeated question by the teacher. Each response will be treated as a feedback to the teacher agent's work from the teacher, and will help the Intelligent Module to improve its heuristic rules as discussed in Section 5.3.5. All questions that are left untouched will be archived into database automatically.

More. The "More" button is a pull down menu that contains a group of less frequently used functions, such as the system configuration tools and the popup quizzes manager. To use the popup quizzes manager, the teacher prepares a set of quizzes in some simple formats, multiple choices for example, before class and organizes them according to the order of the topics covered in the class. During the lecture the teacher agent will send the quiz corresponding to the topic just discussed to student agents. The student agent will display the quiz to the student and start counting time. As the student answers his/her question or the time runs out, the student agent will collect the result and send it back to the teacher agent, where the answer is corrected and the result is saved as part of student's performance. By doing such quick quizzes, agents can help the teacher find out how well the knowledge is transferred and encourage students to come up with better questions.

In order to personalize the help to each individual student's need and accurately assess each student's performance, the teacher agent profiles students from many other aspects too. For example, the teacher agent records every student's login and logout time to make an attendance sheet. If a student came late or left early, the teacher agent is able to transfer the class materials missed by this student, by comparing the timestamps, to the student agent. The student agents also report the students' activities within their buddy groups, as well as asynchronous studies, to the teacher agent for an accurate evaluation. The evaluation information, which is kept absolutely confidential within software agents, is a valuable parameter when student agents first form the buddy groups.

5.4.1.5 The Learning and Student Agent Processes

The student agent always chooses the best route through I-MINDS' layered topological infrastructure for the student to join the virtual classroom. Upon receiving the teacher's Internet location, as shown in Figure 6, the student agent immediately establishes a handshake with the teacher agent, and then the teacher agent uses this handshake message to calculate the Round-Trip Time (RTT) between the two agents. If the RTT is no longer than the predetermined threshold, the teacher agent enrolls this student agent into its multicast group and confirms the enrollment to the student agent. However, if the RTT is larger than the threshold, meaning that the connection between the two agents is slow and may degrade the quality of experience of this student significantly, then the teacher agent informs the student agent to choose a proxy server. At this point, the student agent carries out a real-time test on the list of proxy servers provided by the Manager. It selects the proxy server that has the best RTT time, and subsequently informs the teacher agent of its choice. The teacher agent adds the proxy server into its multicast group, essentially treating it as another destination to send the data to.

By deploying remote proxy servers, we extend I-MINDS' scalability to students whose network has no multicast capability. Because a proxy server serves much fewer students than the teacher server, it is affordable to have one-to-one uni-cast connection with each of its subordinates. Furthermore, the proxy server can provide adaptive quality of service to better accommodate a student's individual network condition. For example, based on the available bandwidth between the proxy server and a student agent, the proxy server may drop some video frames or turn the video steaming off to conserve bandwidth for more essential I-MINDS functionalities.

After joining the virtual classroom, the students will be able to receive the multimedia-enriched lecture by a combination of video, audio, images, and text, and can communicate with the teacher and fellow students in multiple ways. Figure 9 shows a screen snapshot, also captured in our pilot test, of a student taking class using I-MINDS.

The main component at the upper left is the lecture window, where the current presentation (e.g., a slide, a webpage, a document, etc.) the teacher is lecturing on is shown. All handwriting and annotation will be animated in this window with the correct location and proportion. By comparing Figure 8 and Figure 9, it is easy to see that the exact same image on teacher's computer is shown in the lecture window on the student site, minus the question-ranking window that should not be revealed to the students. We call these presentation pages shown in the lecture window the *lecture pages*.

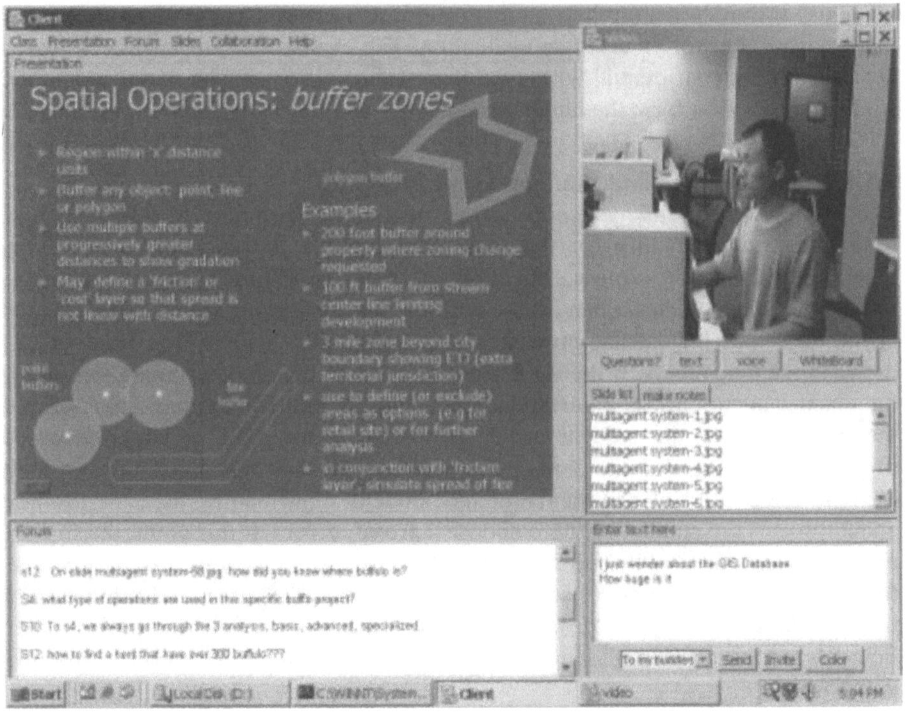

Fig. 9. A Snapshot of a Student's Computer Screen during Our Pilot Test

All lecture pages used in a class are archived locally in to the students' computers. The panel on the middle of right-hand side of the student's window displays a list that enumerates all lecture pages used in the current class session. The student can review any past pages or attach quick notes onto them. When the class session finishes, all lecture pages, together with their corresponding quick notes, will be saved to the directory the student had chosen for asynchronous studies.

At the bottom of the student's application window locates a multiple-color and multiple-format text forum. All incoming messages are displayed at the left with the original color and format of the sender. Outgoing messages are entered at the right. Multiple recipients can be selected from the drop down list.

Sometimes phrasing a question clearly and precisely is no easier than answering it. As an old saying says, *a picture is worth a thousand words.* Thus, associating a question to a specific slide, with which the teacher is very familiar, enables the teacher to understand the context of the question straightaway. In I-MINDS we allow students to superimpose their questions on to the exact spot of the lecture page where the questions arise to simplify the question phrasing process. To ask a question about a specific lecture page, a student double clicks the spot on the page and enters the question in a popup input window. The color of the question can also be selected to make it stand out from the slide's background.

The question and its relative location on the lecture page and its color will be encapsulated by the student agent and sent to the teacher agent. When the teacher reviews this question, the teacher agent will fetch the lecture page from the server's archive and reconstruct the superimposition.

When the question is not related to any lecture page or simple enough to be expressed in text, students can type the question in the forum input panel and send it directly to the teacher by selecting the teacher as the recipient. If the question is too complicated to be articulated in a text message, students can also request a video/audio conversation with the teacher. All requests will be queued by the teacher agent and permissions are granted at the teacher's will.

5.4.1.6 The Collaboration among Students

The hard-to-feel existence of the teacher as well as peer classmates in distance education often makes students feel alone and mentally check out. A collaborative learning environment equipped with effective collaboration tools not only builds strong connections among students, but also enhances participants' deep understanding. In I-MINDS students are encouraged to collaborate in the following ways:

Text Forum. All students can send and receive color-coded, font-specific text messages. Text messages are good to quickly exchange ideas and discuss simple questions. Periodically receiving messages from classmates stimulates students' awareness of each other's existence.

Whiteboard. To assist students in one study group in illustrating problems and ideas, we build a virtual whiteboard in I-MINDS. Anything written or drawn on the virtual whiteboard is broadcast to all other students in the same study group in real-time. The virtual whiteboard pops up on demand, which means students can invoke or turn it off when needed. The student agent will keep track of the virtual whiteboard activities, so that every student's virtual whiteboard is synchronized within the group even if it was turned off. This advanced feature allows anytime "join-ins" of students to the virtual whiteboard, instantly placing newly joined students at the same stage of whiteboard discussion as the other students. A token is available for grab when a student requires exclusive writing permission to stop simultaneously writing on the whiteboard for a finite period of time. A new synchronization and consistency maintenance mechanism based Distributed Shared Object (DSO) [17] in under development to improve the virtual whiteboard sharing experience and performance.

Buddy Group. In a large sized distance education class, it is more practical that students collaborate in small groups. In I-MINDS, students can invite other students to join their collaboration group, or buddy group as we call it. The student agents will help students in forming and refining their buddy group. A student can send messages, via the forum, to other students in the buddy group. Also when a student poses a question to the teacher, the student agent automatically sends the question to the teacher agent and to other students in the buddy group if the student turns this feature on. Because not all questions received by the teacher agent will be addressed by the teacher due to the time constraints, the *buddies* in the buddy group may respond to the questions via the forum, or the virtual whiteboard. The students who are active and helpful in a buddy group will be ranked high in their profiles and preferred by other student agents. The students who have never responded to any messages after a period of time are dropped from the group by the student agents automatically because the student agents profile those students as unsuitable to the buddy group.

5.4.1.7 The Support of Student Agent

In I-MINDS a student communicates with the teacher and other students through the student agent. All data and information coming to the student is processed and presented to the student by the student agent. For example, all lecture pages used in a class and the superimposition of annotation and drawing are reconstructed by the student agent. All out-going messages are encapsulated and sent out by the student agent. The student agent also maintains the synchronization and consistency of the virtual whiteboard. At the same time the student agent archives all lecture pages and messages locally for asynchronous studies.

Besides performing the communication tasks, the student agent also profiles the student and reports the profile to the teacher agent. The student's activities are constantly monitored by the student agent. When the student has been idle, without typing keyboard or moving mouse, for longer than a threshold, the student agent will alarm the student to concentrate on the class. The student agent also tracks the student's participation in the forum and the virtual whiteboard, and the results are shared among the student agents, but concealed from the students, to form and refine buddy groups.

At the beginning of a distance class, students do not know each other well enough to form buddy groups effectively. The student agent will initialize a buddy group for the student based on the profile information queried from the teacher agent and the preferences indicated by the student. Each time a student's profile changes, all student agents will notice and re-evaluate whether to invite that student into their buddy groups. A student can request his/her student agent to invite a particular student to join his/her buddy group. Upon receiving an invitation, the student can accept or reject the request, and further specify accepting or rejecting all succeeding requests.

When a student performs a collaboration activity, such as initializing or participating in a forum or virtual whiteboard discussion, the student agent will inform other student agents identified as buddies. Thus other student agents can perceive and evaluate the goodness of this student. A student agent ranks all buddies in a group based on their responsiveness and helpfulness. Inactive students may be dropped from their buddy groups. And every time a student is dropped from a buddy group, the student agent will report to the teacher agent. When a student is dropped from buddy groups frequently, the teacher agent will inform the teacher to pay special attention to that student.

5.4.2 Asynchronous learning in I-MINDS

With the student agent's help, all class materials are archived locally on to the student's computer. After each class session, the student can explicitly export class lecture pages and the text forum to a specific directory. Otherwise the student agent will automatically save them into the default directory. Through these asynchronous features, I-MINDS help students learn better in terms of organization of notes, review of archived live discussion, and annotation of archived live lectures.

Review the Lecture Pages. Students can type quick notes and stick them onto a slide in class. After class, all slides together with their corresponding quick notes are exported into a selected directory. Then students can review all slides and edit quick notes using the slide reviewer, as shown in Figure 10.

Review the Forum. The archived text forum can be reviewed and edited to better serve the student. Students can insert new text, delete unwanted messages, and change message's colors and sizes. When a message refers to a particular lecture page, it is also possible to include that lecture page into the forum document, as shown in Figure 11. This review process allows a student to organize online discussions into more coherent documents.

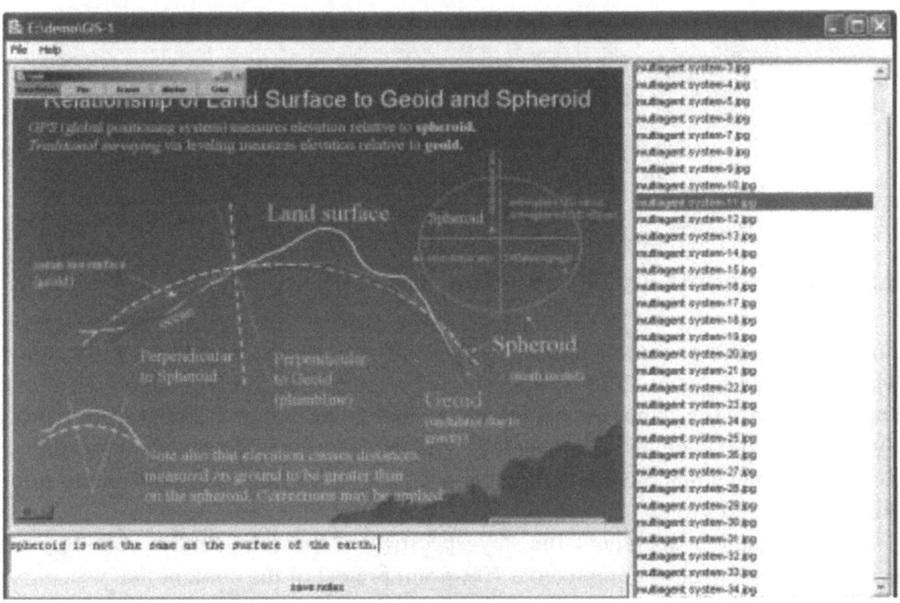

Fig. 10. The Interface of the Lecture Page Reviewer

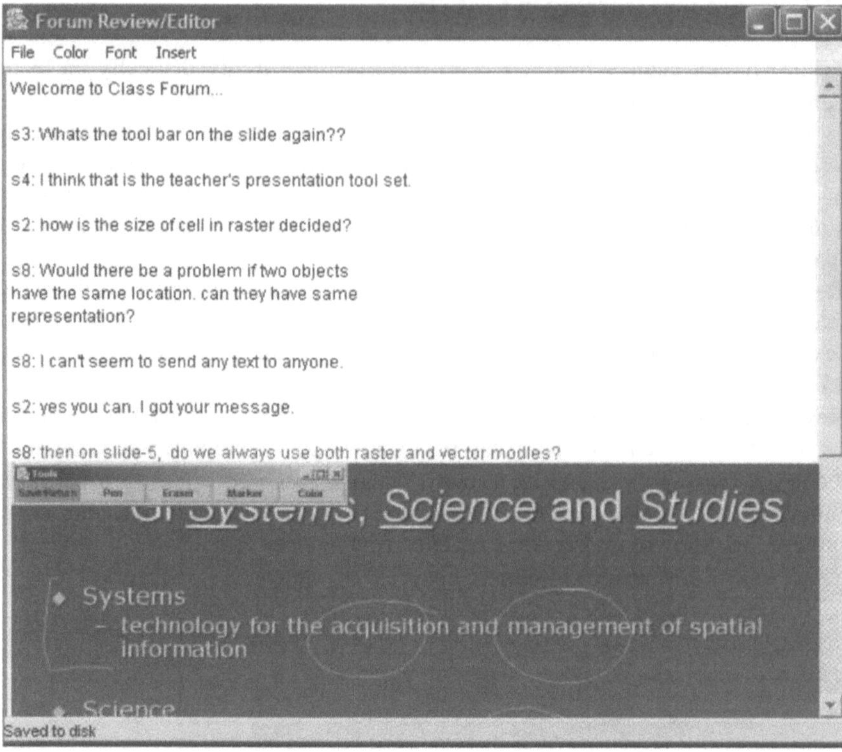

Fig. 11. The Interface of the Forum Reviewer

5.5 Experiments and Evaluation

To determine the potential impact of I-MINDS on student learning, a pilot study was conducted in May 2003 where the tool was used by subjects in a controlled experiment to assess what impact it had on student learning of Global Information Systems (GIS) content. GIS technology can be used for scientific investigations, resource management, and development planning. For further details on our pilot study, please refer to [27].

Tables 1 and 2 document the key specifics of the pilot study. On Day 1, subjects in both groups completed a 109-point pretest of the content that was to be taught during the two sessions. At the conclusion of the class on Day 1 for both groups, the subset of 60 items that related to the content of that class was included on the posttest. After the Day 2 instruction, the subset of 49 items that related to the content of that class constituted the second posttest. Subjects in the control group learned the identical content during each of the two sessions, as did subjects in the experimental group. The difference was that the control group students were in the same room as the instructor. Their class was taught in a very traditional manner with the professor using PowerPoint slides identical to those used for the experimental group to teach the content.

Table 1. Specifics of the Pilot Study.

Specifics of the Pilot Study
All sessions taught by the same instructor
20 undergraduate and graduate students participants
Two groups: Control and Experimental Groups
Each subject received $90 for their participation
This study received IRB Approval.

Table 2. I-MINDS Pilot Study Groups and Events.

Experiment Group	Control Group
Day 1 • I-MINDS Training (30 min.) • Pretest (45 min) • Break (15 min) • Class (60 min) • Distracter Task (5 min) • Posttest (30 min)	**Day 1** • Pretest (45 min) • Break (15 min) • Face-to-Face Class (60 min) • Distracter Task (5 min) • Posttest (30 min)
Day 2 • I-MINDS Training (15 min) • Class (60 min) • Break (15 min) • Distracter Task (5 min) • Posttest (30 min) • Perception Survey (5 min)	**Day 2** • Face-to-Face Class (60 min) • Break (15 min) • Distracter Task (5 min) • Posttest (30 min) • Perception Survey (5 min)

Separate 2 x 2 (group by pretest-posttest) mixed-model analyses of variance were used to determine the effect of using I-MINDS on learning in the experiment (Table 3). Cell means and marginal means for the two groups for the two factors appear below for each day the subjects were tested.

Table 3. Test Results

Test 1			
Group	**Pretest**	**Posttest**	**Marg.Means**
I-MINDS	13.1	33.4	23.25
Control	17.4	41.3	29.35
Marg.Means	15.25	37.5	

Test 2			
Group	**Pretest**	**Posttest**	**Marg.Means**
I-MINDS	12.0	22.2	17.1

Control	15.78	20.67	18.2
Marg.Means	13.79	21.47	

Results for Test 1 revealed a significant main effect for the group factor, $F(1,18) = 5.03$, $p<.05$, with subjects in the control group scoring significantly higher than subjects in the I-MINDS group. This means there was a significant difference found between the two marginal means for this factor. Note that the numbers 1 and 18 refer to the degrees of freedom for the numerator and denominator terms of the analysis. The first number is the number of levels of this factor minus 1, and the second number corresponds to the degrees of freedom for the error term. The main effect for the repeated measures was also significant, $F(1,18) = 131.90$, $p < .01$. The scores on the posttest were much higher than were scores on the pretest. No significant interaction was found, $F(1,18) = .88$, $p> .05$. Results for Test 2 revealed no significant main effect for the group factor, $F(1,17) = .17$, $p>.05$. As was the case for Test 1, the main effect for the repeated measures was significant, $F(1,17) = 17.59$, $p < .05$. The scores on the posttest means collapsed across groups were higher than were scores on the pretest. No significant interaction was found, $F(1,17) = 2.18$, $p> .05$.

Results for the two testing sessions are encouraging. For the initial testing session, it was expected that the control group would either score better than the I-MINDS group or be no different due to the inexperience of both the subjects and the instructor in using the new teaching and learning tool. The training session that preceded the class allowed the subjects to acquire some degree of skill for this first session, but it was hardly sufficient to master the numerous I-MINDS tools.

During the week between Test 1 and Test 2, subjects in the I-MINDS group commented about how they had considered ways to use the software to assist their learning. Although there was a slight difference in the means of the experiment and control groups for Test 2, this difference was not statistically significant (p> .05), and *the amount that the I-MINDS group improved from the pretest to the posttest was nearly twice that of the control group.* This result was very promising as were comments from the subjects in the I-MINDS group related to their comfort level in using the tool.

Comments from the university professor who used I-MINDS in teaching both of the content lessons were also encouraging. He indicated that the teaching tool was very easy to learn and use. He also said that the tool could enhance distance learning, especially by making it possible for building an archive of information that could be accessed "on-demand" by students. The instructor also noted that questions asked of him via I-MINDS tended to be higher quality, reflect a deeper understanding, and demand a richer response than those questions posed during the control sessions.

5.6 Future Work

Our future work for I-MINDS is three-pronged.

First, we plan to bring I-MINDS into deployment and services in the near future. For example, we aim to incorporate I-MINDS into our department-wide Reinventing CS Curriculum, which has been on-going for the past year. Initially, we plan to use I-MINDS in laboratories where paired programming assignments are available. In addition, we want to deploy I-MINDS to support distance education where remote students will be able to use I-MINDS to (a) interact with the instructor and students real-time, (b) interact with the lectures and students in the buddy group off-line through digital archives and digital forums (both available in the current version of I-MINDS), and (c) interact with fellow students in research and project groups such that an instructor can evaluate the roles of the students in a group and assign individual grades more fairly.

Second, we will continue to improve I-MINDS along two fronts. First, we will increase I-MINDS' support in asynchronous learning and incorporate agent's support into asynchronous cooperation. We plan to build a study-group agent, which is similar to teacher agent, to coordinate asynchronous study activities. We also plan to build a question/solution database and an agent that is able to answer questions based on the cases in the database. Thus students can always get help from other online students or the question database. Second, we want to enhance the intelligence of the I-MINDS agents in machine learning—how can each agent learn to adapt to different instructors, students, lectures, and classrooms? To this end, we will also incorporate instructional paradigms in cooperative learning, especially team building. The student agents will adhere to proven techniques and guidelines in team building and also evaluate the quality and progress of a team.

Finally, we see in the future more extensive and comprehensive tests using I-MINDS, primarily in student learning, and secondary in teacher learning. We want to collect data on how students learn with technology, and how students behave with "buddy groups." We want to observe how teachers learn from using I-MINDS as well.

5.7 Conclusions

We have presented an innovative multiagent system to support cooperative learning synchronously and asynchronously among students both in the real classrooms and in distance education. The system, called I-MINDS, consists of a group of intelligent agents. A teacher agent monitors the student activities and helps the teacher manage and better adapt to the class. A student agent, on the other hand, interacts with the teacher agent and other student agents to support cooperative learning activities behind-the-scene for a student. We have described two I-MINDS innovations in (a) agent-federated real-time "buddy group" formation and refinement, and (b) automated ranking of questions and responses. We have reported on the proof-of-concept test of the I-MINDS prototype. The results, as reported, are encouraging. Our future work includes deploying I-MINDS to classrooms, improving I-MINDS technically, and conducting further experiments to learn about technology-supported learning in CS curriculum.

5.8 Acknowledgement

The I-MINDS project is supported in part by NCITE. The authors would like to thank Phanivas Vemuri and Jameela al-Jaroodi, Jeff Lang, and Charles Ansorge for their contributions to the design, implementation, and experimentations of I-MINDS.

References

1. Abdel-Hamid, A., Ghanem, S., Maly, K. and Abdel-Wahab, H. (2001). The Software Architecture of an Interactive Remote Instruction System for Heterogeneous Network Environments. *Proceedings of 6th IEEE Symposium on Computers and Communications*, 2001, Institute of Electrical and Eletronics Engineers, Hammamet, Tunisia. 694-699.
2. Aleven, V., Koedinger, K. R. and Cross, K. (1999). Tutoring Answer Explanation Fosters Learning with Understanding. In S. P. Lajoie & M. Vivet (eds.) *Artificial Intelligence in Education*, Amsterdam: IOS Press, 199-206.
3. André, E. and Rist, T. (2001). Controlling the Behavior of Animated Presentation Agents in the Interface: Scripting versus Instructing, *AI Magazine*, 22(4):53-66.
4. Bonwell, C. C. and Eison, J. A. (1991). Active Learning: Creating Excitement in the Classroom, *ERIC Clearinghouse on Higher Education*, Document No. ED 340 272.
5. Cassell, J. (2001). Embodied Conversational Agents: Representation and Intelligence in User Interfaces, *AI Magazine*, 22(4):67-83.
6. Centra Sofware Company. Centra Symposium 6.0, http://www.centra.com/products/symposium/info.asp [10 August 2003]
7. Gertner, A. S. and VanLehn, K. (2000). ANDES: A Coached Problem-Solving Environment for Physics. In G. Gautheier, C. Frasson, & K. VanLehn (eds.) *Intelligent Tutoring Systems: 5th International Conference, ITS 2000 Montréal, Canada, June 2000 Proceedings* (pp. 133-142), Berlin Heidelberg, New York: Springer-Verlag.
8. Graesser, A. C., VanLehn, K., Rosé, C. P., Jordan, P. W. and Harter, D. (2001). Intelligent Tutoring Sy stems with Conversational Dialogue, *AI Magazine*, 22(4):39-51.
9. Graesser, A. C., Wiemer-Hastings, K., Wiemer-Hastings, P., Kreuz, R. and the Tutoring Research Group (1999). AutoTutor: A Simulation of a Human Tutor, *Journal of Cognitive Systems Research*, 1(1):35-51.
10. Interwise Inc. Interwise E-Learning Solution, http://www.interwise.com/solutions/elearning.asp [10 August 2003]
11. Johnson, D. W., Johnson, R. T. and Smith, K. A. (1991). Cooperative Learning: Increasing College Faculty Instructional Productivity, *ASHE-ERIC Higher Education Report No. 4*, George Washington University.

12. Johnson, W. L., Rickel, J. W. and Lester, J. C. (2000). Animated Pedagogical Agents: Face-to-Face Interaction in Interactive Learning Environments, *International Journal of Artificial Intelligence in Education*, **11**:47-78.

13. Kadiyala, M. and Crynes, B. L. (1998). Where's the Proof? A Review of Literature on Effectiveness of Information Technology in Education, in *Proceedings of 1998 FIE Conference*, 33-37.

14. Koedinger, K. R., Anderson, J. R., Hadley, W. H., & Mark, M. A. (1997). Intelligent Tutoring Goes to School in the Big City. *International Journal of Artificial Intelligence in Education*, 8(1), 30-43.

15. Lesgold, A., Lajoie, S., Bunzo, M. and Eggan, G. (1992). SHERLOCK: A Coached Practice Environment for an Electronics Troubleshooting Job. In J. H. Larkin & R. W. Chabay (eds.) *Computer Assisted Instruction and Intelligent Tutoring Systems* (pp. 201-238), Hillsdale, NJ: Lawrence Erlbaum.

16. Lester, J. C., Callaway, C. B., Grégoire, J. P., Stelling, G. D., Towns, S. G. and Zettlemoyer, L. S. (2002). Animated Pedagogical Agents in Knowledge-Based Learning Environments, in Forbus, K. D. and P. J. Feltovich (2002). (Eds.) *Smart Machines in Education*, Menlo Park, CA: AAAI Press, pp. 269-298.

17. Liu, X., Jiang, H. and Soh, L.-K. (2004), A Distributed Shared Object Model Based on a Hierarchical Consistency Protocol, In *Proceedings of 4th IEEE/ACM International Symposium on Cluster Computing and the Grid*, Chicago, Illinois, USA, April 19-22, 2004.

18. Liu, X., Zhang, X., Al-Jaroodi, J., Vemuri, P., Jiang, H. and Soh, L.-K. (2003). I-MINDS: An Application of Multiagent System Intelligence to On-Line Education, in *Proceedings of the IEEE International Conference on Systems, Man, and Cybernetics*, Washington, D.C., October, 4864-4871.

19. Liu, X., Zhang, X., Soh, L.-K., Al-Jaroodi, J. and Jiang, H. (2003). A Distributed, Multiagent Infrastructure for Real-Time, Virtual Classrooms, accepted to *International Conference on Computers in Education (ICCE2003)*, Hong Kong, China.

20. Marsella, S. and Johnson, W. L. (1998). An Intelligent Assistant for Team Training in Dynamic Multi-Agent Virtual Worlds, in *Proceedings of Intelligent Tutoring Systems, 4th International Conference*, 464-473.

21. Raines, T., Tambe, M. and Marsella, S. (2000). Automated Assistants to Aid Humans in Understanding Team Behaviors, in *Proceedings of the Fourth International Conference on Autonomous Agents*, 419-426.

22. Rickel, J. and Johnson, W. L. (1999). Animated Agents for Procedural Training in Virtual Reality: Perception, Cognition, and Motor Control, *Applied Artificial Intelligence*, **13**(4):343-382.

23. Sen, S. and Weiss, G. (1999). Learning in Multiagent Systems, in Weiss, G. (ed.), *Multiagent Systems: A Modern Approach to Distributed Artificial Intelligence*, MIT Press, pp. 259-298.

24. Sheppard, S. D., Reamon, D., Friedlander, L., Kerns, C., Leifer, L., Marincovich, M. and Toye, G. (1998). Assessment of Technology-Assisted Learning in Higher Education: It Requires New Thinking by Universities and Colleges, *Proc. 1998 FIE Conference*, 141-145.

25. Shih, T. K., Distance Education Technologies: Current Trends and Software Systems, *Proceeding of the First International Symposium on Cyber Worlds (CW'02).*

26. Silberman, M. (1996). Active Learning: 101 Strategies to Teach Any Subject, Allyn & Bacon.

27. Soh, L.-K., Jiang, H. and Ansorge, C. (2004). Agent-Based Cooperative Learning: A Proof-of-Concept Experiment, *Proceedings of the 35th Technical Symposium on Computer Science Education (SIGCSE'2004)*, March 3-7, Norfolk, VA, pp. 368-372.

28. Soh, L.-K., Liu, X., Zhang, X., Al-Jaroodi, J., Jiang, H., Vemuri, P. (2003). I-MINDS: An Agent-Oriented Information System for Applications in Education. *Proceedings of AAMAS'03 Workshop on Agent-Oriented Information Systems (AOIS),* 2003, Melbourne, Australia. 2-8.

29. Soh, L.-K., Liu, X., Zhang, X., Al-Jaroodi, J., Jiang, H. and Vermuri, P. (2004). I-MINDS: An Agent-Oriented Information System for Applications in Education, *Lecture Notes in Artificial Intelligence (LNAI 3030) Series Agent-Oriented Information Systems*, Springer-Verlag, pp.16-31.

30. Virtual Ink Corporation. Mimio Classroom, http://www.mimio.com/meet/classroom [10 August 2003].

31. Weiss, G. (ed.) (1999). *Multiagent Systems: A Modern Approach to Distributed Artificial Intelligence*, MIT Press.

32. Wooldridge, M. and Jennings, N. R. (1995). Intelligent Agents: Theory and Practice, *The Knowledge Engineering Review*, 10(2):115-152.

33. Woolf, B. P., Beck, J., Eliot, C. and Stern, M. (2002). Growth and Maturity of Intelligent Tutoring Systems: A Status Report, in Forbus, K. D. and Feltovich, P. J. (2002). (Eds.) *Smart Machines in Education*, Menlo Park, CA: AAAI Press, pp. 99-144.

6. Intelligent Agents to Improve Adaptivity in A Web-Based Learning Environment

C. I. Peña[1], J. L. Marzo[2], and J. Ll. de la Rosa[2]

[1] Information Technologies Division, Industrial University of Santander, Bucaramanga, Colombia

[2] Electronics, Automatics and Computer Science Department, University of Girona, Spain

This chapter focuses on the use of intelligent agents in on-line learning environments in which educational organizations can equip students with lifelong learning skills for today's society. In this scenario, and for this case, a web agent can be thought of as a software package with the potential to improve the guidance provided to the user through personalized contents considering learning styles and cognitive states. Agents proposed here, will provide the students with personal assistants than can help them to carry out learning activities according to their learning styles and knowledge level. The student's progress is tracked and his/her motivation during learning is also taken into consideration. The agent's environment is built by means of a multiagent architecture (MASPLANG) designed to support adaptivity (adaptive presentation and adaptive navigation) in a hypermedia education system (USD) used for distance learning on the web. A distinguished feature of the proposed approach is the ability to build a hybrid student model beginning with a student stereotype model which considers the student's learning style and it is gradually modified as the overlay model is built from information acquired from the student's interaction (subjective likes) within the learning environment.

6.1 Introduction

The proposed system called MASPLANG[1] [14] is built using a two-level agent architecture (*Information* and *Assistant* agents) as shown in Figure 1.

[1] MASPLANG: MultiAgent System PLANG. PLANG belongs from the Spanish acronym "PLAtaforma de Nueva Generación"

C.I. Peña, J.L. Marzo, and J.Ll. de la Rosa: *Intelligent Agents to Improve Adaptivity in A Web-Based Learning Environment*, StudFuzz **178**, 141–170 (2005)
www.springerlink.com

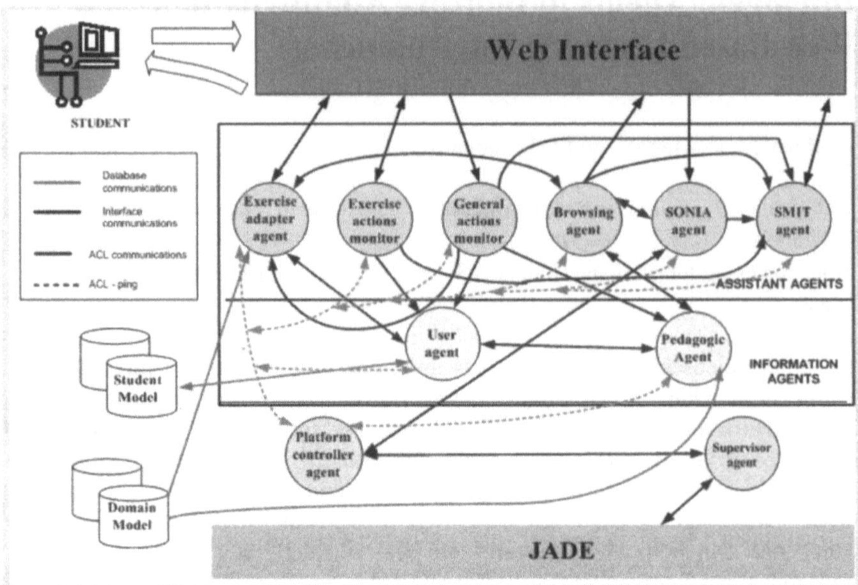

Fig. 1. MASPLANG two-level agent architecture for adaptive education

The *assistant* agents are designed to assist students as they work with the didactic material arranged for the course. Such assistance consists of registering the student actions (by means of the *Monitor* agents) to identify patterns that can be used for personalizing the presentation of the learning content and the navigation tools the students use to move through the contents (by means of the *Browsing* agent) and to adapt exercises (by means of the *Exercise Adapter* agent) for self-assessment to the student's knowledge level or preferences. To make the student feel comfortable and motivated when he/she carries out the learning activities, an animated, life-like character (the *SMIT* agent) [2] has been designed to display the reinforcement information and the programmed alert messages (by means of the *SONIA* agent).

There are two *Information* agents. The first is the *User* agent designed to maintain the student model, and the second is the *Pedagogic* agent which evaluates the pedagogic decision rules that are embedded in the pedagogic model of the course. The *Information* agents are very close to the system databases (students' learning activities dossier and domain model).

In this context, the MASPLANG agents are designed with the following properties [13] taken into account:

- *Reactivity*: agents need to maintain a continuous relationship with their environment and respond to the changes that happen in it.
- *Interactivity*: agents need to interact with each other in order to achieve the goals.
- *Autonomy*: agents need to know when and how to carry out the tasks assigned to them.

- *Proactivity*: agents have goals or explicit objectives (i.e. to find didactic contents in graphic media, to select structural navigation tools, etc.) and need to act accordingly and in an autonomous manner to achieve them.
- *Learning*: the User agent learns from student interactions in order to adapt the learning environment to the student model (learning profile and student knowledge state).

The agency and personalization model of this system follows the behavior shown in Figure 2: Students (rectangles) interact with an environment (USD[2] [1] platform) through agents (circles) that represent them. The agents have a double function: *interacting* with each other and with the habitat on behalf of the student and *filtering* the information (type and style of didactic contents, navigation tools and navigation techniques) that the students receive from other agents and the habitat. The agents are individuals (each student has his/her own agent) and they all have knowledge about the objectives and learning styles of the students they represent; they are also capable of learning from interactions with the environment.

In the following sections the operation of this architecture shortly together with its conceptual model and the methodology for its development and implementation is described.

6.2 Client-Server Architecture

The MASPLANG is built on JADE[3], a FIPA compliant multiagent system [3], using Java, JavaScript, Flash, PhP, JSP, HTML and XML languages at different stages of the agents' programming.

The USD teaching and learning working space (the virtual desktop) was programmed using HTML, CSS Style Sheets, JavaScript and ActionScript (Macromedia Flash) languages. Its interface was divided into the following four frames to facilitate the working area definition for each assistant agent (see Figure 3):

- The frame on the right (number 4 in Figure 3) displays all the working windows of the environment (tool interfaces, learning contents, exercises, etc.).
- The frame on the bottom (number 3 in Figure 3) displays the general tool bar of the environment.
- All assistant agents are invisible in frame 1, where a JADE container (a Java applet) is loaded. The *Exercise adapter* and the *SMIT* agents have a visible representation in the navigation tool bar (at the top of the screen in frame 1).
- *SONIA* agent has a visible representation in frame 2. In the background, the monitor agents' register student actions in all frames (except the *exercise action* monitor which goes into action only during events happening inside an exercise opened in the right-hand frame).

---

[2] USD: belongs from the Spanish acronym "Unidades de Soporte a la Docencia"

[3] JADE: Java Agent Development Framework

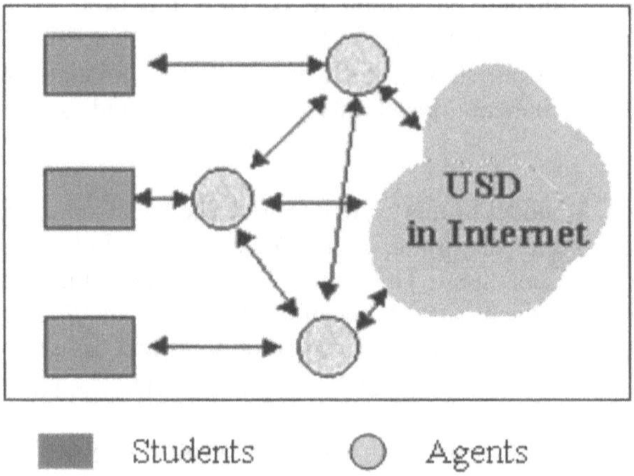

Fig. 2. MASPLANG model of agency and personalization

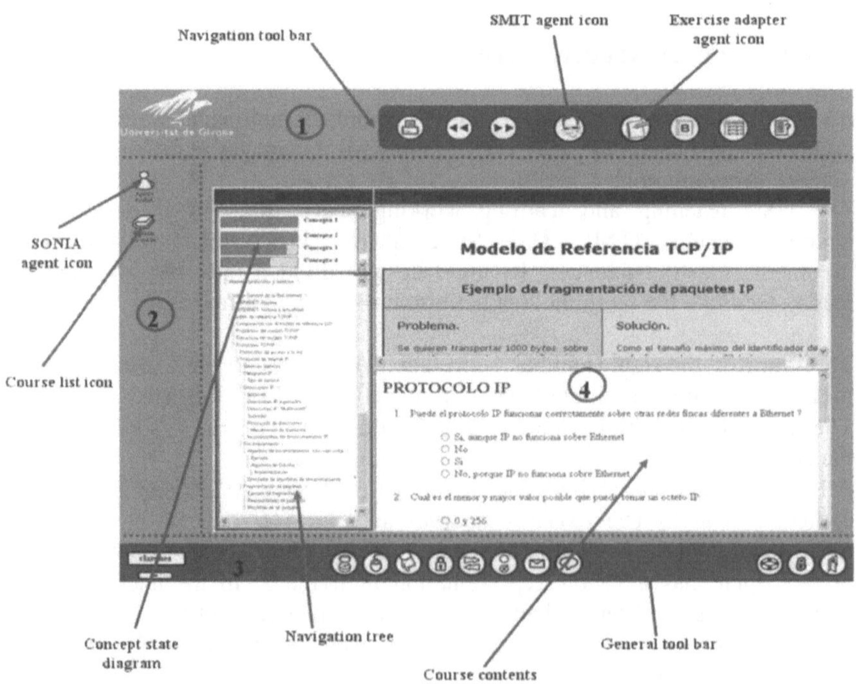

Fig. 3. Aspect of the USD working space

6.3 Working Scenario

The working scenario of the MASPLANG is defined by the type of users and the type of contents offered. As the working environment is an adaptive hypermedia system for education, the users are classified as *professors*, who prepare and set up the teaching content for adaptive learning and *students* who carry out the learning activities in a personalized way.

In order to create the initial student learning profile, the system asks the student to answer the ILS[4] questionnaire [4]. This questionnaire consists of a set of questions of a psychological nature whose goal is to determine the student's wants, habits and reactions that will act as a guide, in part, for personalizing the content and the learning environment. The student model is built by taking into account this learning profile and the student knowledge state obtained by analyzing the student actions.

The teachers' interaction with the system may be summarized as follows:

Teachers build the teaching content based on a set of HTML pages that comprise the theoretical definitions (declarative knowledge – **what** should be taught) using different instructional designs and media formats (to match the FSLSM learning styles for the information *processing* and *reception* dimensions). Subsequently, using the teacher's tools available in the environment, they proceed to define **how** these contents should be taught (domain model building), in which case they build the concept structure and the relationships between the concept elements. Finally, this knowledge is stored in the system database and the student carries out the learning activities in a pleasant and assisted environment through the personalized user interface.

The following figure (Fig 4) shows this working scenario.

[4] ILS: Index of Learning Styles. A diagnostic instrument of the FSLSM [6] learning style model adopted for this study

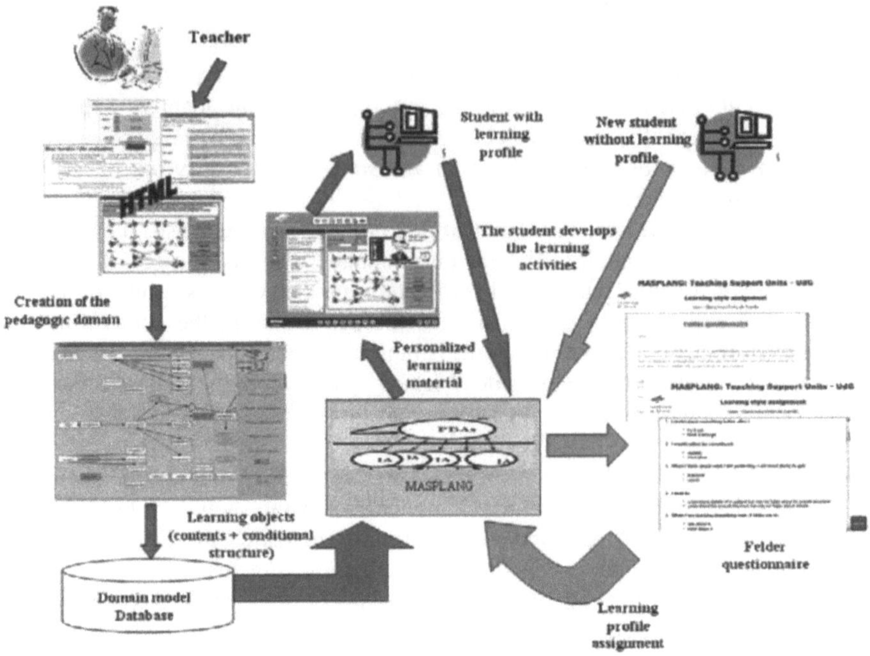

Fig. 4. MASPLANG working scenario

6.4 Agent Services

6.4.1Services of the *Monitor* Agents

The aim of the *Monitor* agents is to register student activity from the student learning environment as the learning tasks are carried out. These monitoring tasks consist of registering the student mouse-clicks on relevant buttons of the entire working desk during a learning session - when he/she studies a lesson, completes exercises or enters the system for the first time. Therefore, the student model is updated by the *User* agent performance which processes the collected information.

The gathered activity also allows the fine-tuning of the student learning profile by the agents of the HabitatPro™ environment (a tool for personalization and market prospecting developed by Agents Inspired Technologies Corporation [5]).

6.4.2 Services of the *Exercise Adapter* Agent

The aim of the *Exercise Adapter* agent is the construction of suitable exercises for a student learning session. This process is carried out with the following two features taken into account:

- The student's preferences, in which case it is the student who configures the topics and the types of questions that he/she wants to answer (configured exercise).
- The student's knowledge level, in which case it is the agent who selects the topics and the types of the questions that the student should answer in a given moment (adapted exercise).

An exercise is in fact a group of multiple choice questions. Each one of these questions is associated with a topic and a level of difficulty according to the domain model structure. There are three levels of difficulty and they are described as 1-*easy*, 2-*normal* and 3-*difficult*.

There are two types of exercises in a lesson:

- *Mandatory* exercises. These are represented as prerequisite nodes in the navigation map. In this case, it is the teacher who determines the general characteristics of the exercise that the agent should create for the student, for example, the number of questions to complete, their level of difficulty, the number of possible attempts at the exercise that the student is allowed, the total time that the student may spend on the exercise, etc.
- *Optional* exercises or self-assessment exercises. In this case, the student may determine the general characteristics of the exercise to complete. The student may also request an adapted exercise from the *Exercise Adapter* agent according to his/her level of knowledge.

6.4.3 Services of the *User* Agent

The student model represents the computer system's belief about the learner's knowledge. In order to allow instruction to be individually tailored, it is first necessary to capture the student's understanding of the subject. With this information, the difficulty of the material and any necessary remediation can be controlled within the instructional system. Building a student model involves defining:

- The "*who*", or the degree of specialization in determining who is modeled and what the learner history is;
- The "*what*", or the goals, plans, attitudes, capabilities, knowledge and beliefs of the learner;
- "*how*" the model is to be acquired and maintained;
- And "*when*" to give assistance to the learner, to provide feedback or to interpret learner behavior.

In maintaining the student model, the factors that need to be considered include the fact that students do not perform consistently, they forget information randomly and then exhibit large leaps in understanding. The student model, which is the essential component when offering individualized learning in e-learning systems, is the one that builds and maintains the system's understanding of the student.

In the context of the MASPLANG, it is the *User* agent that builds and maintains the student model - taking into consideration the domain model (domain and pedagogical knowledge) and the student performance. The *Monitor* agents collect all the information concerning the student performance for the *User* agent and the *Pedagogic* and the *Exercise Adapter* agents consult the *User* agent for information about the student model in order to adapt the contents and the navigation paths for a particular student.

6.4.4 Services of the *Pedagogic* Agent

In the context of the MASPLANG, it is the *Pedagogic* agent which defines the navigation paths and the content that a student may study progressively in a learning session according to the student model (learning profile and knowledge state) and the structure of the domain. To carry out these adaptive tasks, the *Pedagogic* agent evaluates the decision rules of the pedagogic domain, requesting suitable information about the student model from the *User* agent. The information that the student receives is presented by the *Browsing* agent through a personalized interface with ergonomic navigation tools.

The process of building the navigation tree or the concept diagram is carried out by means of a constructor (at the implementation level) which loads data from the domain model database and builds a tree data structure or a bar diagram respectively, after evaluating the pedagogic decision rules embedded in the pedagogic model using the information from the student model, which is managed by the *User* agent.

Figure 5 shows the diagram of this process. In the navigation tree structure it is important to notice the construction of strong and light links which delimit the suitable path and nodes that the student may follow at that moment.

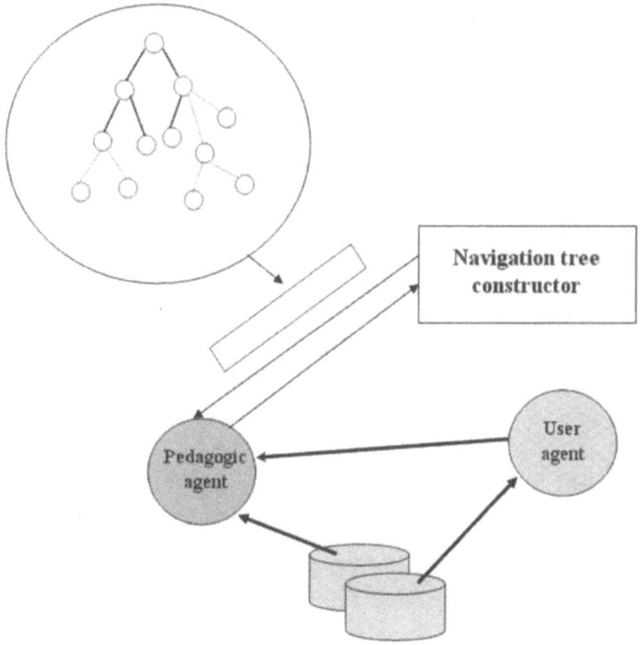

Fig. 5. Information flow and processes that allow the navigation tree to be built

6.4.5 Services of the *Browsing* Agent

The *Browsing* agent is an assistant agent which creates, in the student interface, the navigation structure of the learning content (HTML pages) adapted to the student learning profile and to the student level of knowledge. The adaptive navigation techniques [16], such as *hidden link*, *direct guidance* and *link annotation* as well as the selection of suitable navigation tools, are applied to assist the student in navigating the contents in a personalized way.

As it operates, the *Browsing* agent communicates with:

- The *Pedagogic* agent (which builds and maintains the navigation tree and the concept state diagram according to the student model) in order to refresh the information to be presented;
- The *SONIA* agent in order to indicate which nodes have particular information for review, if the student has programmed it to provide such alerts.
- The *Exercise Adapter* agent if the lesson has exercises assigned.
- The *SMIT* agent in order to send the information that should be represented to the student in a user-friendly interface (to motivate or to reinforce behaviors).

At the implementation level, the work developed by the *Browsing* agent consists of the dynamic construction of two HTML pages (by means of JSP programs) that enable the navigation tree and the concept state diagram (built by the *Pedagogic* agent) in the course interface. Figure 6, shows, a representation of this process in the USD environment.

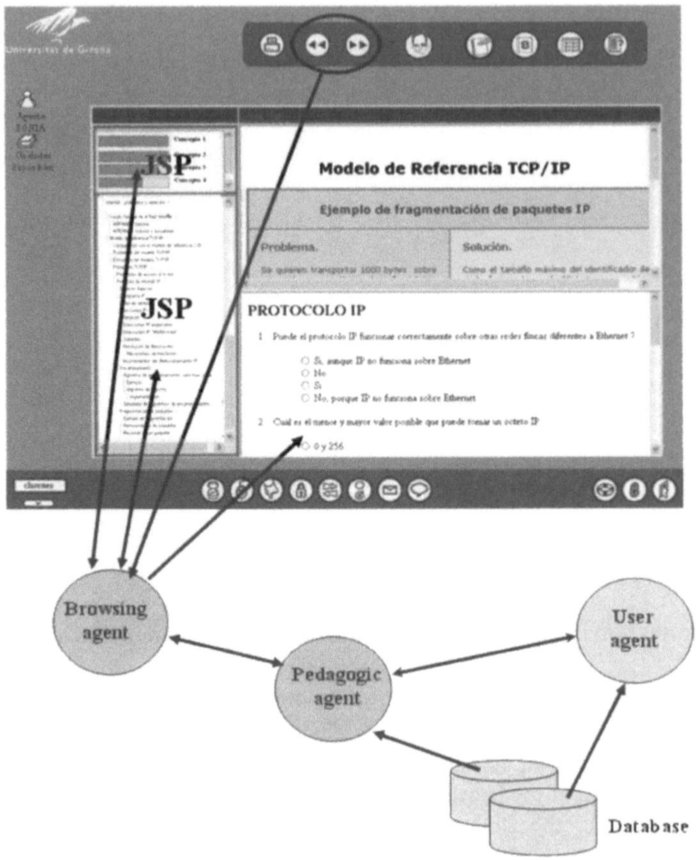

Fig. 6. Working space of the *Browsing* agent in the USD environment

6.4.6 Services of the *SONIA* Agent

SONIA (Student Oriented Network Interface Agent) is a very simple agent. It was designed to perform tasks that the users (teachers or students) may program. Some of these tasks are:

- To inform the student when a specific classmate comes on-line.
- To suggest looking at the bibliographical references in some sections of the lesson.

- To suggest doing the interactive exercises proposed when the student gets to particular sections of the lesson.
- To alert the student if he/she has gone beyond a specific time of study.
- To provide the student with personalized messages (reminders, scheduled events, etc.) at specified times.
- In the special case of a professor's message, to get the attention of students currently connected to the system so that they could revise some specific sections of the lesson, solve a particular problem or enter the chat room to carry out an on-line discussion.

To achieve these goals, *SONIA* agent works cooperatively with the *Controller*, *Browsing* and *SMIT* agents as follows:

- With the *Browsing* agent, by requesting alerts about the existence of bibliographical references to review or exercises to carry out.
- With the *Controller* agent, by requesting the information on certain system events (i.e. alarm clock, user's login, broadcast message, etc).
- With the *SMIT* agent, by reporting the messages that should be presented to the student if the tasks have been completed.

6.4.7 Services of the *SMIT* Agent

SMIT (Synthetic Multimedia Interactive Tutor) is a synthetic agent. It is introduced in the environment using an animated interface (anthropomorphous). Its goal is to show the student the messages (i.e. warnings, motivation, feedback, etc.) coming from other agents in the environment. (e.g., to interrupt the student with a warning message from the *SONIA* agent). Each message representation demands the selection of certain animations and body movements to define the *SMIT* behavior in any particular situation. The aim of using this agent is to "humanize' the learning environment and to make it user-friendlier and closer to the student.

Figures 7, 8 and 9 show aspects of this agent design and performance.

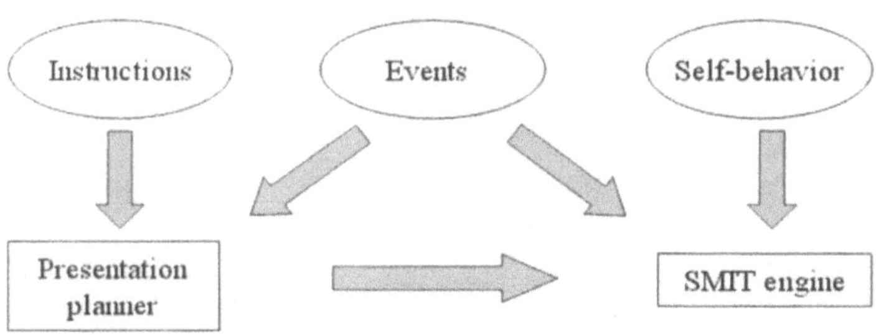

Fig. 7. Presentation planner structure of the *SMIT* agent (adapted from [15])

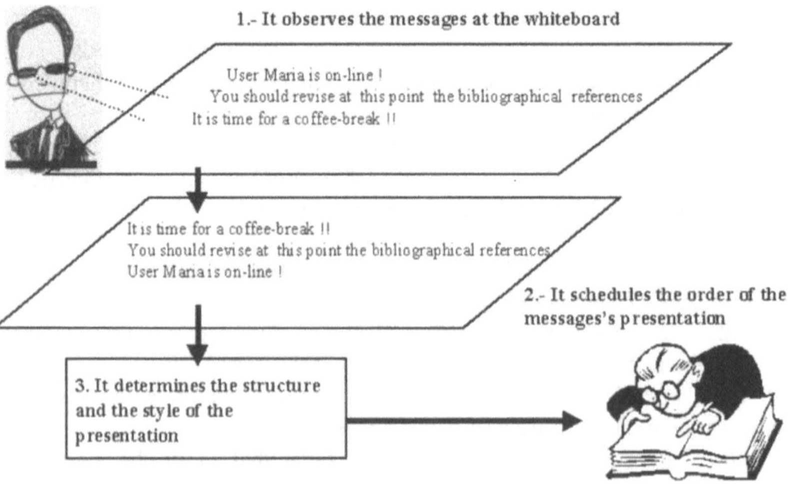

Fig. 8. General representation of the *SMIT* agent operation

Fig. 9. Aspect of the *SMIT* agent representing some messages in the learning environment

6.5 Conceptual Model

The conceptual model of the USD adaptive hypermedia system based on the MASPLANG agent approach, is composed basically of three models: the *domain model*, which determines the concepts to be taught and their interrelationships in order to provide a global structure of the domain concerned (knowledge domain); the *student model*, which allows the different features of the student (such as expertise, knowledge, preferences, objectives, etc.) to be considered in the learning process; and the *interaction model*, which encapsulates the adaptive engine that provides adaptive presentation and adaptive navigation by means of supervising the student interaction (the global operation of this model was described in section 2.).

The next formulation represents an abstract model of the MASPLANG performance, taking into account the conceptual model described above:

IS: $$S \leftarrow LP_{ls} \quad (1)$$

D_{ls}: $$C * O \rightarrow \sum_{i=1}^{n} HTML \quad (2)$$

ES: $$SM_S \leftarrow LP_{ls} * K_S(D_{ls}) \quad (3)$$

Where:

(1) Means that at the beginning of the session, a *learning profile* based on *learning styles* (***LP_{ls}***) is assigned to the student ***S***. The first value of this learning profile is obtained by evaluating the ILS questionnaire given to students. Later on, after collecting a representative number of student actions, this profile is fine-tuned through a case-based reasoning process carried out by the agents of the HabitatPro™ tool.

(2) Defines the knowledge domain model based on learning styles. It consists of a set of concepts (**C**) with an organization structure (***O***). In the end, what the student receives from this domain model is a set of HMTL pages.

Finally, (3) means that at the end of the learning session, the student model (***SM_S***) is updated with the student learning profile (***LP_{ls}***) and the student knowledge state (***K_S***) taken into consideration.

6.5.1 Domain Model

The domain model (***D_{ls}***), as part of an adaptive hypermedia system for education, represents both the knowledge about a particular domain that will be transmitted to the student and the way of presenting that information (rules defined in a pedagogic model). The domain model knowledge and its structure determine the contents of the tutorial interaction, together with the structure that governs the adaptive instruction.

The domain model of the MASPLANG is *declarative* and its knowledge is represented by means of a conceptual map (see Figure 10) whose structure takes into account the *static (*the ***what***) and *evolutionary (*the ***how***) focus.

From the ***static*** point of view, the teaching concepts are represented by means of a conceptual network structured using different taxonomies. Each node corresponds to a domain concept and it is disaggregated in other nodes using *class-subclass* relationships (i.e. tree like structure). The resultant conceptual network is a static representation of the knowledge in the teaching domain (i.e., ***what*** will be taught).

From the ***evolutionary*** point of view, the conceptual network is structured using relationships to describe the pedagogic rules needed to select the contents and/or determine their sequencing. In this study, conceptual (i.e. *property* relationships, such as "X is part of Y") and *procedural* relationships are considered. The procedural relationships are used to determine the *order* in which the concept nodes should be learned or the *decisions* that should be evaluated to reach any instructional objective (i.e. if condition A is true then the student may study nodes 1.1 and 1.2 of the *Concept* 1). This structure corresponds to the didactic organization of the domain (i.e., ***how*** the concepts will be taught).

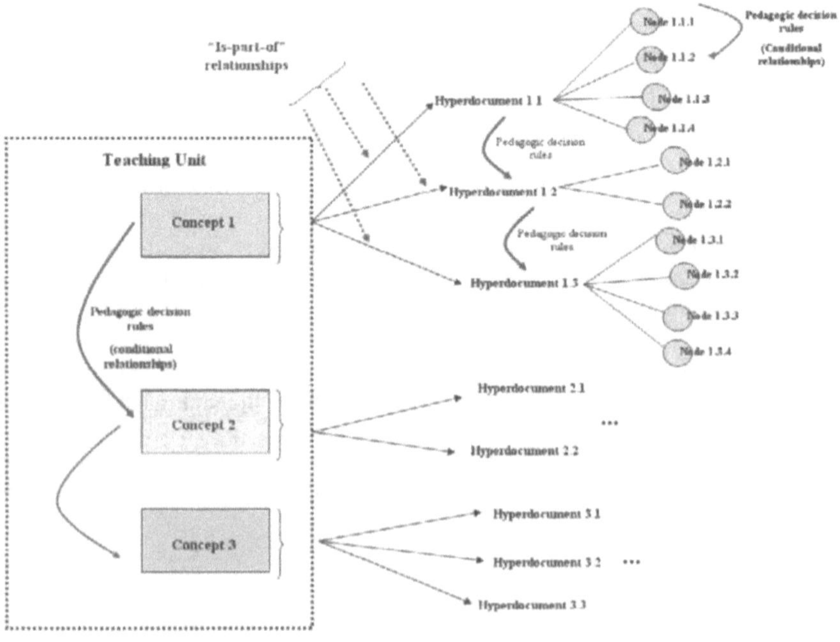

Fig. 10. Example of the MASPLANG domain model organization

6.5.1.1 Contents Type for Learning Styles

Supporting quality teaching and learning has been one of the critical issues in distance education. In distance learning scenarios, one of the key things to consider is how the student feels about the educational material. This raises several critical issues concerning learning styles and dynamic pedagogic material adapted to particular student preferences.

Selecting the learning style model is crucial to the development of an effective adaptive hypermedia course that addresses different learning style preferences. For this study, the FSLSM learning style model was adopted since it has been well tested in web-delivered courseware for Engineering and Computer Science education (our field of interest). The experiences of Carver [7] using this model, have demonstrated that students are empowered to learn using their own unique learning style instead of being forced to learn according to the instructor's point of view.

The FSLSM model offers four dichotomous dimensions that identify eight learning styles that may be associated to moderate or strong tendencies, as described below:

- The *Processing* dimension involves *active/reflective* learning styles.
  - *Active learners* tend to acquire knowledge by doing something. They like to try out things, and bounce other people's ideas around. In addition, they feel comfortable with group work.
  - *Reflective learners* process the information introspectively, and normally they think things through before trying them out. Generally they prefer to work alone or in pairs.
- The *Perception* dimension involves *sensitive/intuitive* learning styles.
  - *Sensitive learners* learn better when the information presented includes facts and procedures.
  - *Intuitive learners* tend to be imaginative, prefer interpretations and concepts, like variety in their work, do not mind complexity and get bored soon with too much detail and repetition.
- The *Input* dimension involves *visual/verbal* learning styles.
  - *Visual learners* get more information from visual sources, such as pictures, videos, diagrams, graphs, schematics and demonstrations.
  - *Verbal learners* are comfortable with written and spoken communications.
- The *Progress* dimension involves *sequential/global* learning styles.
  - *Sequential learners* prefer to approach knowledge in small steps of connected chunks (blocks of information).
  - *Global Learners* like to approach information in apparently unconnected chunks and achieve understanding in large holistic leaps, connecting all the chunks intuitively.

Following the experimental work applied by Carver and using a similar approach that takes advantage of the versatility offered by the teaching tools of the MASPLANG environment, the teaching contents and the navigation tools to match learning styles have been adapted. Adapting some traditional instructional strategies and building the learning objects by means of HTML pages (since MASPLANG teachers have worked well developing contents in this format) which have subjects embedded in different media format, Table 1 offers a useful distribution of criteria for selecting the right instructional strategies, instructional complementary materials, interactive and assessment elements and navigation tools for adaptive presentation and adaptive navigation.

6.5.1.2 Domain Model Representation

The MASPLANG domain model is represented by a semantic graph based on a set of concepts to teach. Each concept is considered as a *basic learning unit* with its own properties (i.e. associated learning style, required level of knowledge, requisites, etc.). As was shown in Figure 10, a concept in the semantic graph is disaggregated in hyperdocuments and these in turn are disaggregated in nodes. For this study, the node contents were prepared by teachers using HTML pages while ensuring that they matched learning styles. The relationships between concepts and hyperdocuments and between hyperdocuments and nodes are represented by links in the graph.

The graph may have different types of nodes and links as shown in Figure 11.

Table 1. Hipermedia course componentes for MASPLANG considering FSLSM learning styles

a. Instructional Strategy
b. Media format
c. Navigation tool

a.→	Lesson Objectives	Case studies	Lectures	Knowledge nucleus	Conceptual maps	Synthesis
Active				√		
Reflective	√	√	√		√	
Sensing		√			√	
Intuitive	√				√	
Visual		√			√	√
Verbal	√		√		√	
Sequential					√	
Global	√					√

b.→	Slideshow		Media clips			Lineal
	Text	Multimedia	Graphics	Digital movies	Audio	Texts
Active						√
Reflective	√					√
Sensing		√	√	√	√	√
Intuitive	√	√	√	√	√	√
Visual		√	√	√		
Verbal	√				√	√
Sequential	√	√		√	√	√
Global			√	√		

c.→	Punctuals			Structurals		Collaborative work		
	Arrows	Printings	On line help	General vision maps	Filters	Chat	Forum	e-mail
Active	√	√		√	√	√	√	√
Reflective	√	√	√	√	√			√
Sensing	√	√	√	√	√	√	√	√
Intuitive	√	√	√	√	√	√	√	√
Visual	√	√	√	√	√	√	√	√
Verbal	√	√	√	√	√	√	√	√
Sequential	√	√	√			√	√	√
Global				√	√	√	√	√

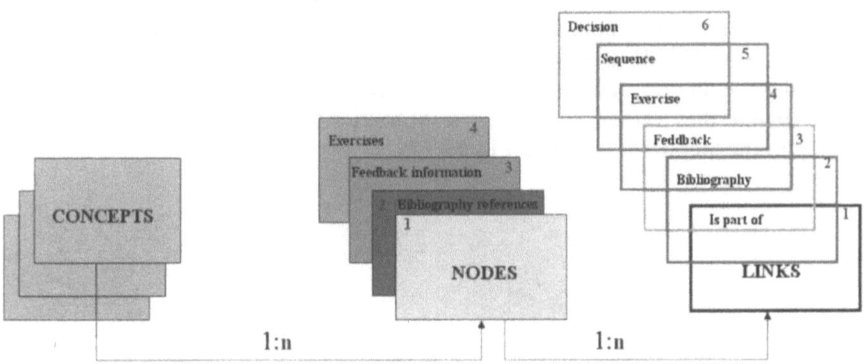

Fig. 11. Nodes and link types in the semantic graph

The main nodes should contain:

- *Basic information* (corresponding to theoretical explanations – node 1 in Figure 11) or information for *enforcing student assessment* (nodes that correspond to exercises – node 4 in Figure 11).
- Additionally, there are two optional types of nodes that may be associated to each main node of theoretical explanations (these nodes will be available when the main nodes are being studied). These are:
  - The *bibliography* nodes (node 2 in Figure 11) that provide a review of the bibliography, and
  - The exercise nodes (node 4 in Figure 11) that provide *exercises for student self assessment* (notice that exercise nodes may be of the *main* or *optional* types depending on how they are linked in the graph).
- There are six types of links as follows:

- The *is-part-of* link - link 1 in Figure 11 - which connects the nodes that are part of a hyperdocument used to explain a concept or part of a concept.
- The *bibliography* link - link 2 in Figure 11 - which points to an optional bibliography node (node 2 in the set of nodes). This link may be enabled from any of the main nodes of the graph.
- The *feedback* link - link 3 in Figure 11 - which points to a feedback information node (node 3 in the set of nodes). This link may only be enabled in a *decision* link.
- The *exercise* link - link 4 in Figure 11 - which points to an optional exercise node (node 4 in the set of nodes). This link may be enabled from any of the main nodes of the graph.
- The *sequence* link - link 5 in Figure 11 - which points to the next node to be visited, thus establishing a mandatory sequence.
- The *decision* link - link 6 in Figure 11 - which allows the user to visit the next node once a particular condition is satisfied.

The information concerning exercises is the only one that is not prepared directly by teachers using HTML pages. Instead, before making the graph, teachers should build (and save in the database) a global exercise skeleton with questions and answers (using the Exercise editor – a MASPLANG teaching tool) organized according to the hierarchy shown in Figure 12.

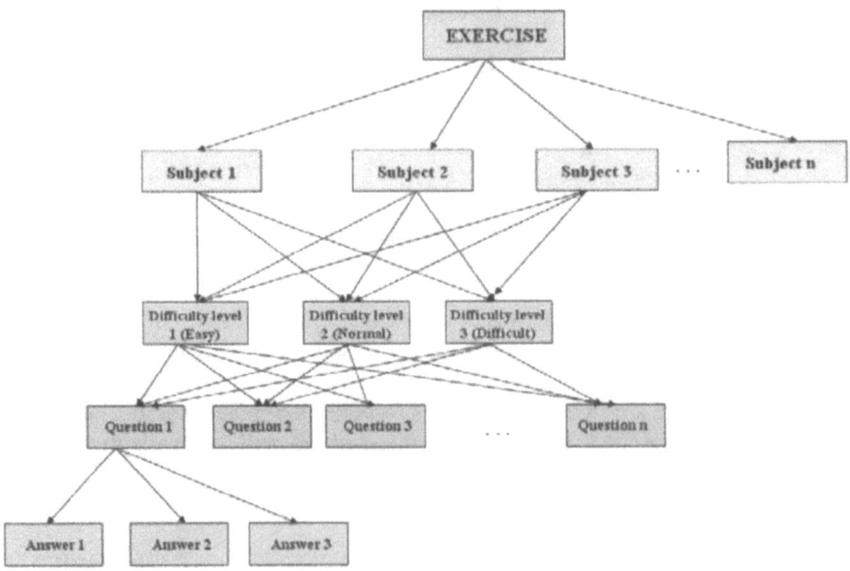

Fig. 12. Exercise hierarchy organization

The exercise nodes referred in the semantic graph are built during the student learning session by the *Exercise adapter* agent, in HTML format, using instances from the global exercise skeleton. At this point, the agent (which appears as an icon in the navigation tool bar) allows the student to decide if he/she wants to configure an instance of the exercise or to leave the agent to adapt the exercise according to the student knowledge state (applying principles of the Gagné theory [8]). The following rules show aspects of this mechanism for exercise adaptation [9]:

- ***If*** the student has failed the last exercise (i.e. achieved less than 5) ***then*** propose an exercise with the same level of difficulty (based on the *repetition principle*; repetition of the same schema under different appearance improves learning).
- ***If*** the student has passed the last exercise with a score above 7, ***then*** propose another one with a higher level of difficulty (based on the *logic order principle*).

In order to build this domain model, the MASPLANG offers teachers the *Teaching Units Editor* which allows the semantic graph to be represented graphically. This way of working (i.e. physically drawing the graph) with the facilities afforded by the editor (ergonomic and easy-to-use environment) was found to be highly acceptable by the MASPLANG teachers [10].

Figure 13 shows an example of the semantic graph built for the course "*Study of the TCP/IP protocol*s" used as a prototype in the MASPLANG experimentation and evaluation.

6.5.2 Student Model

A student model defines a knowledge base that establishes: the learning characteristics of a student; the knowledge that he/she has about the domain; the didactic material that he/she has used to learn; the history of the learning sessions that he/she has carried out; etc. This information is used by other components of the system to achieve a more efficient process of instruction which is better adapted to the student.

In this section, we describe how the student is modeled in MASPLANG by means of the *User* agent within the multiagent architecture.

6.5.2.1 Student Modeling Technique

Two elements are taken into account when modeling the student in MASPLANG: the *student model* knowledge base and the *User* agent (i.e., the student manager – an agent based on knowledge).

The student learning characteristics are established in the knowledge base. MASPLANG uses a *hybrid* model, a combination of an *overlay* model [11] and an *inferred* model, to represent the student knowledge about the domain. This model is in turn divided into two more conceptual models: one is *permanent* and the other is *temporal*.

The *permanent model* contains information concerned with data about the student's personal characteristics and his/her learning profile; the knowledge that he/she has about the domain; the didactic material that he/she has used to learn and the history of the learning sessions carried out (common actions, history of exercises, etc.). This model is available during the whole instruction process and is updated session by session.

The first data acquired for the *learning profile* comes from the ILS questionnaire (a task carried out by the *general monitor agent* of the multiagent architecture). Later, it is fine-tuned by analyzing the student's interactions with the system, using a procedure described in the next section.

The *knowledge about the domain* is the knowledge that the student has acquired through the learning process and how this knowledge was acquired. The particular structure of the domain is modified to include new attributes to control the acquisition characteristics (the navigation through the graph is adapted to the student knowledge state). These attributes are: the acquired-level to indicate the level of knowledge that the student has about a concept and the concepts that the student has learnt.

The *didactic material that the student has used to learn* (basic contents and exercises) identifies the material used by the system to present the learning content or to assess it. This information is used by the *pedagogic* agent of the MASPLANG multiagent architecture to make a suitable choice of the content that the student should learn at any particular time. The *exercise adapter* agent uses the information concerned with the exercises already solved by the student to adapt exercises to the student knowledge state.

In the student model, the information about the development of the instruction process is also considered. Data about the last session summarizes the learning session events that took place in the last session. This information is important for setting up certain elements to be represented in the next session.

The *history of the whole instructional process* is represented by a set of actions commonly carried out, which nodes were visited along with the time spent on each visit and information about the student development when solving exercises (a list of the exercises that the student has made and the way they were solved). This information is available through the learning statistics button in the general tool bar of the learning environment.

The *temporal model* makes sense only for the current session. Its data is managed for the *User* agent who at the end of the session updates the *permanent model* with relevant information that should alter the student knowledge state.

The *User* agent is the student manager. Its job is mainly to identify the student's objectives, updating the student model and fine-tuning the learning profile.

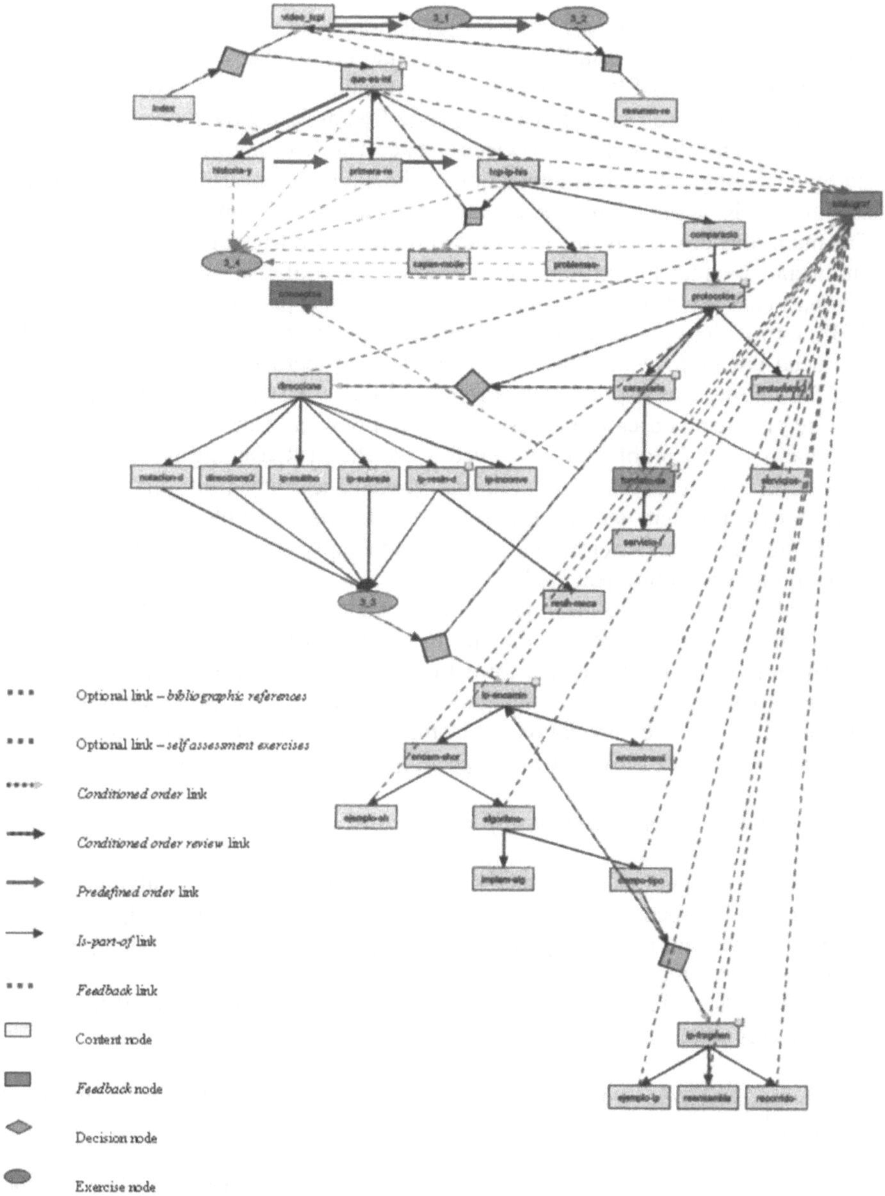

Fig. 13. Semantic Graph of a MASPLANG Course

6.5.2.2 Tuning the Student Learning Profile

An interactive system, in order to adapt its behavior to the needs of the user, must be capable of dynamically building a representation of the user's interests and characteristics. The MASPLANG *user* agent models the student according to the student learning profile (with the learning style first assigned by evaluating the ILS questionnaire) and the student knowledge state. The learning profile is then fine-tuned by analyzing the student interaction in the environment using the Case Based Reasoning (CBR) approach implemented in the HabitatPro™ tool.

The main idea of CBR is to solve a new problem by retrieving a previous similar situation and by reusing information and knowledge of that situation. Finding a similar past case and reusing its solution can solve a new problem in a new situation. In CBR terminology, a *case* usually denotes a *problem situation*. A previously experienced situation, which has been captured and learned in a way that can be reused for solving future problems, is referred to as a past case, a previous case, a store case or a retained case. Correspondingly, a new case or unsolved case is the description of a new problem to be solved. Case-Based Reasoning is – in effect – a cyclic and integrated process of solving a problem, learning from this experience, solving a new problem and so on. If two problems resemble each other, their solutions are similar and therefore it is possible to apply an adaptation of the former solution to the current problem [12].

In the MASPLANG, the information of the student model allows the adaptation of the learning environment to the needs of the individual students. Adaptive presentation (concerning the media format of the contents and the best subject for a particular situation) and adaptive navigation (concerning the appropriate navigation techniques and suitable navigation tools for comfortable navigation through the subjects) are the adaptive hypermedia aspects applied in this approach.

System performance for student profiling

In order to get an agent to act correctly in the best interests of the individual that it represents, it is indispensable that it incorporates knowledge, in some way, of his/her likes, preferences and personality. This knowledge is, in general, extremely diffuse and contradictory (subjective) and therefore difficult to represent and manage. The MASPLANG User agent obtains this subjective knowledge by means of images of the student it is representing.

Characterization of the subjective particularities of the teaching units by means of attribute-value pairs

The agent's personalization techniques involved in the student profiling by the HabitatPro use the concept of *attribute-value* pairs, which are used widely in Artificial Intelligence to represent knowledge. The attribute applicable to a teaching unit or to a student is equivalent to a property or a characteristic.

For the didactic contents, some of the following possible attributes are considered:

- The media format for content presentation (i.e. graphic, text, hypertext, audio, etc).
- The type of interactivity that the content offers (i.e. sensitive maps, simulations, and exercises)
- The instructional strategy used to explain the situations (i.e. objectives, summaries, examples, synthesis, and lectures).

Each attribute may have one value among a group of possible values. If a group of elements is defined, the elements to which certain sets of attributes are applicable can be identified. In the MASPLANG, three groups of adaptive elements are considered:

- The didactic contents,
- The navigation techniques and
- The navigation tools.

The values that the attributes can take are generally of a subjective nature, because the meaning of each one depends on the person that uses or defines it (one student may learn a teaching unit better in a graphic format than one in a text format) and therefore sophisticated techniques are needed for their manipulation. In this case, adding more values of this type does not make sense. The algorithms used to deal with these values in the system are based on AI techniques, such as Case-Based Reasoning and Fuzzy Logic.

The image of an element is defined as a set of pair attribute-values that characterize that element. For instance, the image of the *Introduction to Computers* teaching unit for an *active* student (as defined by the FSLSM model) may contain the following attribute-value pairs:

Media format: hypertext
Interactivity elements: simulations
Type of Contents: conceptual maps
Presentation style: hypertext

While for a *sensitive* student it might be:

Media format: video
Interactivity elements: simulations
Type of Contents: examples
Presentation style: video-clips

Using the concept of element images, it is possible to define the new concept of *distance* between elements using Fuzzy Logic techniques. This distance allows us to obtain, from the images of two elements, a numerical value that represents the degree of similarity existing between them. This distance is a function:

$$d_p : P \times P \rightarrow R \quad (4)$$

Where P is the set of element images and R is the set of real numbers.

Characterization of the subjective particularities of the students by means of the triple elements attribute-value-weight

A set of *attribute-value* pairs related to a student can reflect his/her preferences with respect to the teaching units and the learning environment. For instance, an *active* student (FSLSM model category) could be characterized for a specific teaching unit with the following set of pairs:

{*Media format*/hypertext,
Interactivity element/simulations,
Type of Content/conceptual maps,
Presentation style/hypertext's links,
Navigation techniques/direct guidance,
Navigation tools/arrows,
Collaborative work/forum}

Representing student preferences by means of an *attribute-value* pair is not efficient because it does not take into account the intensity of the preferences or the importance that the student gives to each of the attributes. To solve this problem, a weighting is associated to each attribute and to each student. Each student will give to each attribute his/her own weighting, which will indicate the importance the attribute has for him/her when assigning a degree of preference to an element.

The set of weights that can be associated to the attributes is configurable. In this system the following weights are used:

W= {Indifferent, Less Important, Medium Important, Important, Very Important and Necessary}

The variable values for these are {0, 1, 2, 4, 8 and 1000}. Here in it can be observed that *W* will always contain two special constants: *Indifferent* with value zero and *Necessary* with a very big value.

From this point, the image of the student in a group is defined as a set of three elements: *attribute/value/weight* which encodes the student's preferences and the importance given to the values of the attributes for the elements of the group.

By introducing the weight parameter, the following examples show the characteristics of the Teaching Units preferred by a student with an *active* learning style and by a student with a *reflexive* learning style:

***Teaching Unit for an active student*:**

{Media format/hypertext/very important
Interactivity elements/simulations/important
Type of contents/conceptual maps/very important
Presentation style/hypertext links/important
Navigation techniques/direct guidance/important
Navigation tools/backward and forward arrows/necessary

Collaborative work/forum/important}

An *active* learner tries to acquire the knowledge by doing; he/she likes to work in groups and is comfortable navigating the contents by means of the direct guidance navigation technique, whereby the backward and forward navigation arrows are necessary.

Teaching Unit for a reflexive student

{Media format/graphic/very important
Interactivity elements/simulations/very important
Type of contents/conceptual maps/indifferent
Presentation style/hypertext links/very important
Navigation techniques/direct guidance/indifferent
Navigation tools/backward and forward arrows/indifferent
Collaborative work/forum/less important}

A *reflexive* learner processes the information introspectively, acquires the knowledge better by means of graphical contents, thinks a lot before acting and prefers to work alone or in pairs.

By extending the concept of *distance* between products introduced in (4), and generalizing it to include the *weights* of the attributes, two new concepts are included:

1. *The distance between a student and a Teaching Unit, which is defined by*:

$$d: P \times C \rightarrow R \qquad (5)$$

Where *P* is the set of teaching Unit images and *C* is the set of student images.

Given an image *c* of a student and an image *p* of a Teaching Unit, we get the function:

$$d\,(p, c) \qquad (6)$$

This function considers, simultaneously, all the attributes used in the two images, along with their values and weights in order to return a numerical value representing the distance between the student and the element. Therefore function *d* takes the subjective information representing the student and the teaching unit images to obtain a concrete numerical measure of the affinity between them.

2. *The distance between two students, which is defined by*:

$$d_c : C \times C \rightarrow R \qquad (7)$$

Where *C* is the student image set.

Given two student images, *c* and *d,* we get the function:

$$d\,(c\,,d) \tag{8}$$

This function considers, simultaneously, all the attributes used in the two images along with their values and their weights, in order to return a numerical value that represents the distance between the two students.

In a similar way to the function (6), function (7) uses the subjective information representing two student images to obtain a concrete numerical measure of the affinity between them.

The applications of the functions (6) and (8) are immediate. With them, it will be possible to tell a student about the didactic materials that will work best for him/her, to put students into groups according to related preferences or to use the information studied by a student in a teaching unit to promote the same didactic materials for other students of similar learning styles (*collaborative filtering*). Another type of application of great importance for these functions is the analysis and the prospecting of new teaching units.

Upgrading student images

As was mentioned previously, an essential characteristic of intelligent agents consists of being capable of learning from its interactions with other agents and with the environment. In this case, where the agents are incorporating knowledge on student preferences and personality, the learning process will consist of continuously fine-tuning the student images so that, gradually, they reflect a more faithful likeness.

To this end, each time the student carries out activities in a Teaching Unit, his/her image will be updated and adjusted to the new situation. Not only will the system maintain something like an average value for each one of the attributes, but it will also automatically upgrade the weightings, so that the attributes for which the student always chooses the same values (or nearly the same) will have higher weightings, while the attributes for which the student has no marked preference for a particular value or range of values will have a lower weightings.

The magnitude of the image upgrade will be controlled according to different factors:

- *The demonstrated student interest*: If a student more or less carries out all of the learning activities proposed in a teaching unit that has certain attributes, the system will assume that he/she likes this unit and therefore the magnitude of his/her image upgrade for that specific unit, should be bigger than for another teaching unit that presents the same learning contents with different attributes but which the student has never gone into or, if he/she has gone into it, has only carried out a minimum of the learning activities proposed there.
- *The quality of the student interaction*: the magnitude of the student image upgrade will be related to the quality or quantity of interactions that he/she has had within the learning environment.
- *The type of teaching unit:* the system can more faithfully upgrade the image of a student if the teaching unit is rich in content with diverse learning activities.

- *The time:* it has to be remembered that people's preferences change with time. Therefore the upgrade of the student image will be more representative *if a moderate time has passed* since the last upgrade.
- *The student's preferences:* the system should be sufficiently open to allow the student to change his/her image according to his/her own preferences.

6.6 Conclusions

In this chapter, the logical and technological solution proposed for the development of an adaptive e-learning system based on intelligent agents was presented. The main motivation in proposing the MASPLANG architecture and methodology was the need to offer students the didactic material best suited to their individual learning profile. This is achieved in combination with a user-friendly style, assisted and customized environment provided by the former implementation of a Course Management System named USD.

In achieving the proposed objectives of adaptivity, Agent Technology has been successfully applied. Working with agents has been highly interesting. Although the proposed architecture is quite complex, the parallelization of the work allowed by such technology was crucial in achieving the desired functionality. The systematic design of the ontology facilitates the overall process. It is also remarkable that the addition of the agent gives a more comfortable feeling for users, in particular the anthropomorphic aspect of the *SMIT* agent when presenting messages.

The conceptual model of the proposed adaptive hypermedia system is based on a standard architecture; however, the domain, student and interaction models have been designed and added on the basis of previous experience using the USD platform.

There is a prototype MASPLANG system available for some experimentation with reduced groups. This allows us to check some of the desired features of the MASPLANG methodology, in particular the behavior of the agent architecture and its "influence" on this new way of working on the platform, for both teachers and students. The collaborative work of the agent system eventually provided the expected adaptivity.

In conclusion, the main features of the MASPLANG can be summarized as follows:

- The system makes it easier for teachers to design, publish and manage their courses. A set of high level, flexible and ergonomic tools are provided for creating the domain model, to check on student progress and to communicate with them.
- Assistant agents can be set up at the client side (browser). This avoids unnecessary communication with the agents, at the server side, that can result in a degradation of the performance of the overall system.
- Students can access their study material through an adaptable and adaptive environment. In particular, I would like to point out the role of the exercise adapter and the SMIT agents which are closer to the users assisting them during their learning process.

- Experimentation results show that teachers tend to re-use the electronic materials prepared previously in text formats to build the new e-learning units, especially units built exclusively around theoretical contents. The learning activities prepared exclusively for distance students used the collaborative work tools a great deal. The teaching environment was easy to use for those professors who were used to working with computers to produce material or who had the opportunity to practice using the teaching tools sufficiently. The effectiveness of the tools offered to the professors to follow up students' performance was appreciated, as the enthusiasm demonstrated by teachers involved in this new teaching modality.

With regard to the students' behavior and attitude, the main conclusions are:

- Students felt comfortable developing the learning activities proposed in the teaching units concerned;
- They carried-out the learning activities, motivated to achieve a good result in the final exams when using self-assessment materials.
- A slight tendency towards behaviorism was detected in the students' learning processes; students still used only those USD platform resources that were strictly necessary. The guidance of teachers or tutors still has a great influence on student activity.
- Some students do not accept this new modality of education at all, preferring to study using printed materials.
- The students wanted to use more interactive materials with a combination of different media formats.

The schema in Figure 14 summarizes the methods and techniques involved. From this experience, we may conclude that the proposed solution is viable for the e-learning community, who expect a personalized and assisted education with a touch of "humanity".

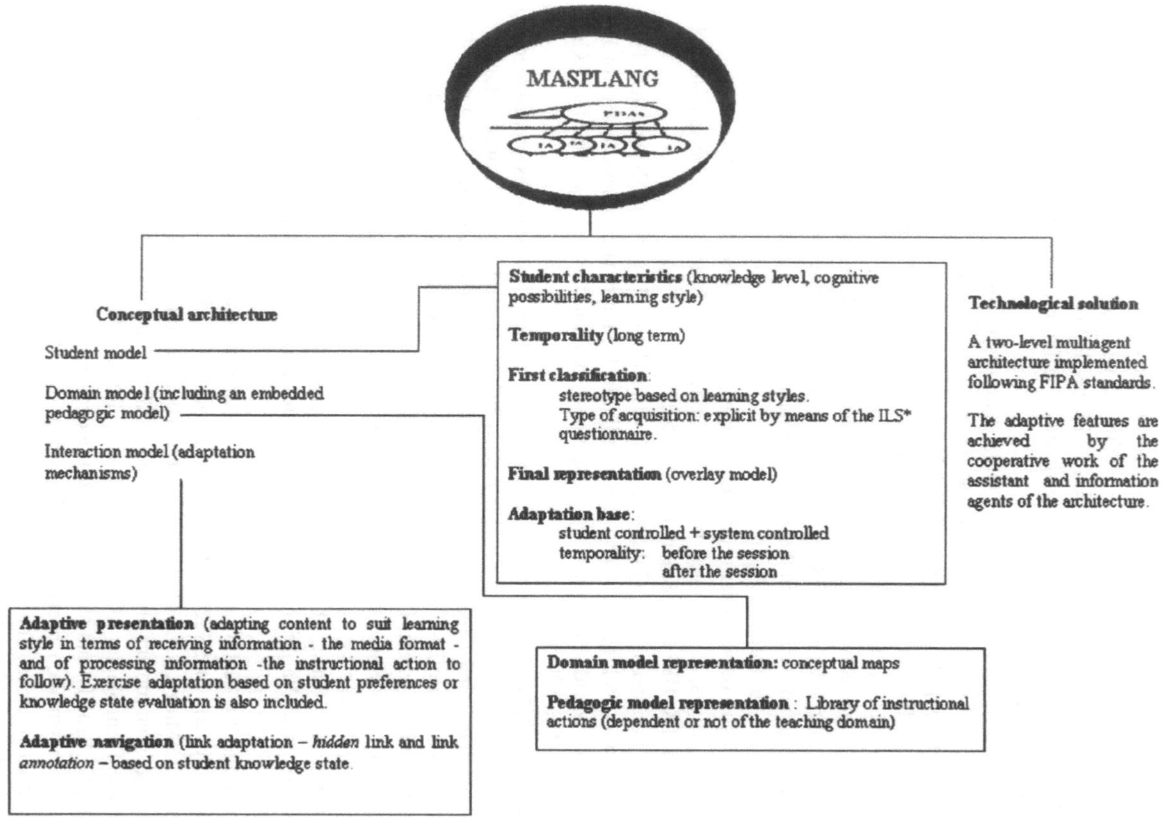

Fig. 14. Methods and techniques for MASPLANG development

References

1. Fabregat, R., Marzo, J.L., Peña, C.I. (2000): Teaching Support Units, Kluwer Academic Publishers: Computers and Education in the 21st Century. 163-174.
2. Aguilar, M., Peña, C. I., Fabregat, R. (2002): SMIT, diseño e implementación de un agente sintético de presentación para las Unidades de Soporte a la Docencia del PLAN-G, VIII Jornadas de Enseñanza Universitaria de la Informática, JENUI 2002, Cáceres (España), 313-320.
3. Java Agent Development Framework. http://sharon.cselt.it/projects/jade.
4. Diagnostic instrument of the FSLSM learning style model. http://www2.ncsu.edu/unity/lockers/users/f/felder/public/ILSdir/ilsweb.html.
5. Habitat-Pro™ Environment, Agents Inspired Technologies Corporation, University of Girona, Girona, Spain, (2001), http://www.agentsinspired.com.
6. Felder, R. M. and Silverman, L. K. (1988): Learning and Teaching Styles in Engineering Education, Eng. Education, 78(7), 674-681. The paper is preceded by a 2002 preface that states and explains changes in the model that have been made since 1988.
7. Carver, C. A., Howard, R. A. and Lane, W. D. (1999): Addressing Different Learning Styles Through Course Hypermedia, IEEE Transactions on Education, 42(1) 33-38.
8. Gagne, R., Briggs, L. & Wager, W. (1992): Principles of Instructional Design (4th Ed.). Fort Worth, TX: HBJ College Publishers, 165-179.
9. Pérez, T., Lopistéguy, P., Gutiérrez, J. & Usandizaga, I. (1995): HyperTutor: From hypermedia to intelligent adaptive hypermedia, InH. Maurer (Eds.), Proceedings of ED-MEDIA'95, World conference on educational multimedia and hypermedia . Graz, Austria: AACE, 529-534.
10. Marzo, J.L., Peña, C.I., Mantilla, C., Carrillo, L. (2002): Evaluating distributed learning at the University of Girona, First GALECIA (Group for Advanced Learning Environments using Communication and Information Aids) workshop, IE2002, Vigo (Spain), Noviembre 20-22, http://eia.udg.es/~atm/bcds/pdf/udg-paper-galecia.pdf.
11. Carr, B. and Goldstein, I. (1977): Overlays: A theory of modeling for computer aided instruction, (AI Memo 406), Cambridge, MA: Massachusetts Institute of Technology, AI Laboratory, Context.
12. Aamodt, A., Plaza, E. (1994): Case-Based Reasoning: Foundational Issues, Methodological Variations, and System Approaches, AI Communications, IOS Press, Vol. 7:1, 39-59.
13. Wooldridge, M. (2000): Intelligent Agents: Introduction, 2nd European Agent Systems Summer School, EASSS2000, Course 1.1, August 14 – 18. Saarbrucken, Germany, 1-67.
14. Peña, C. I., Marzo, J. L., De la Rosa, J. Ll. (2002): Intelligent Agents in a Teaching and Learning Environment on the Web, 2ond IEEE International Conference on Advanced Learning Technologies (ICALT2002), Kazan (Russia), September 9-12 21-27.
15. André, E., Rist, T., and Müller, J. (1998) Integrating reactive and scripted behaviors in a life-like presentation agents. In K.P. Sycara and M. Wooldridge (Eds.), Proc. of the Second International Conference on Autonomous Agents, 261-268.
16. Brusilovsky, P. (1996): Methods and techniques of adaptive hypermedia, Journal of User Modeling and User Adapted Interaction, 6, (2-3), 87-129.

7. Intelligent Virtual Teaching

Goran Šimic, Dragan Gaševic, Zoran Jeremic, and Vladan Devedžic

FON – School of Business Administration, University of Belgrade, POB 52, Jove Ilica 154, 11000 Belgrade, Serbia and Montenegro

There are two major groups of existed adaptive education systems: Adaptive Hypermedia (AH) and Intelligent Tutoring Systems (ITS). While the atention of the ITS is directed on the pedagogoical activities, and knowledge and learner modeling, the LMS are enpowered to accomplish different administraiton tasks: management of knowledge, course and student groups. The interoperabillity between different ITS and LMS requires the standardized data representation and technology which support those standard formats. The Semantic Web could be seen as an opportunity to solve the problems of interoperabillity.

7.1 Introduction

Using current Internet technology to support learning in the classroom is recently becoming much easier and much more feasible than it used to be. If a network of computers or workstations is available in a classroom (the same is on the global network), it is easy to install and use Apache, Tomcat, or another Web server. It can easily distribute HTML pages generated statically or dynamically by an educational application. Client computers/workstations should only have an Internet browser. Hardware and software requirements for the client machines are minimal.

Two groups of the adaptive education systems are the most frequently used on the Web. Those are Adaptive Hypermedia (AH) and Intelligent Tutoring Systems (ITSs). The AH systems are focused on non-linear and adaptable structure of the educational materials [6]. AH systems provide to the user easy navigation, referencing and global view to the content. Also, they provide presentational adaptation techniques (the conditional or stretch text, variants of pages and fragments, and frames linked to the concepts).

Both of them (AHS and ITS) are narrow focused on the specific area of one domain. While the AH systems have compact system design with high coupled components [5], the ITSs have high-level modularity. ITSs provide the user (student) oriented design and much more pedagogical knowledge implemented in the system. Today there are many AH and ITS stand-alone systems that are used for similar educational tasks. The same knowledge is developed at the same time on the different places. This is the typically waste of domain experts' time. Therefore these systems are usually expensive and can not be used without license, payment or/and registering.

G. Šimic et al.: *Intelligent Virtual Teaching*, StudFuzz **178**, 171–202 (2005)
www.springerlink.com

The learning management systems (LMSs) are much more successful in Web-enhanced education which is related to the number of users. LMSs are integrated systems that support a number of teachers' and students' needs. Teachers can use a LMS to develop Web-based course materials and tests, to communicate with students and to monitor their progress. Students can use it for learning and collaboration. Although the adaptive education systems perform every function much better than an LMS, today there is a complete dominance of LMSs over adaptive educational systems.

LMSs provide a teacher to compose their courses from newly created and also existed learning units (so called learning objects - LO). These objects are modeled and described by standard structure and metadata. This means that LOs would be reused in many courses and for different purposes. The standardization means that an LO could be found on the different locations on the Web, and semantically can be connected with the number educational structures at the same time.

Intelligent LMSs (ILMSs) are the bridge between the modern approach to Web-based education (based on learning management systems) and powerful but underused intelligent tutoring and adaptive hypermedia technologies [6]. The ITS reusing of an supported domain in great deal of courses can be realized by the well-described knowledge. This knowledge has to be expressed in a precise, machine-interpretable form which enables the interoperable application components to process LO data, as well as on the syntactic and semantic level [9]. The Semantic Web, a recent Web community effort [2], is a promising technology for improving semantic interoperability of LOs [35]. The main parts of the Semantic Web are domain ontologies. Those ontologies should provide a formal description for a shared domain conceptualization [19]. As the new Web generation [6], the Semantic Web has better conditions for composing and reusing learning materials. The Semantic Web could be seen as an opportunity to enhance the metadata associated to learning materials, expanding the possibilities of current e-Learning specifications and standards [14].

We are going to try in this chapter to explain the main characteristics of the ILMSs, and to show our approach to create an ILMS called Multitutor, as a Semantic Web enabled system. In the next section we are going to give an overview of ILMSs and identify their shortcomings regarding interoperability. Section three explains the Multitutor architecture. Section four shows the Multitutor implementation in detail. In section five we show three courses developed in Multitutor: Code Tutor for teaching radio-communications, Design Pattern Tutor and a Petri net teaching system. Section six discusses how could we benefit of using Semantic Web technologies in development of e-learning systems.

7.2 ILMS – General Concepts and Applications

Nowadays, there are many different ITSs and LMSs. But the educational needs are not yet satisfied. There is no interoperability between these systems. The main problem is that every kind of data on the Web is poorly structured. The existing structures do not have a standardized format. In the last years the community tries to define the ontology of different kinds of knowledge [29]. The great task is that the existing systems accept those standards and modify their data and applications accordingly to standard representations and interfaces.

The ILMS structure is based on both structures - ITSs and LMSs. As with ITSs, in the ILMS there are modeling and representing relevant aspects of knowledge. This means that it contains the knowledge about a student, the domain, the pedagogy and the communication.

The general concepts that support the above knowledge aspects are implemented as components of ITS architecture. There are five basic ITS modules (Figure 1): *student model, domain knowledge, pedagogical module, expert model and communication model* [1]. The communication model is an interface for a student - system communication. This module provides the possibility that more users can be in the session with the system at the same time. Also the communication model dispatches appropriate learning contents to individual users.

The pedagogical module is a tutoring part of an ITS. Different learning strategies and teaching tactics would be implemented in this module. The pedagogical module is responsible for the decisions about every individual student. This module profiles the student and determinates the student model stereotype. During the student sessions, the pedagogical module measures the students' skills and knowledge, and updates the student model. The system changes its behavior according to the students' skills and knowledge.

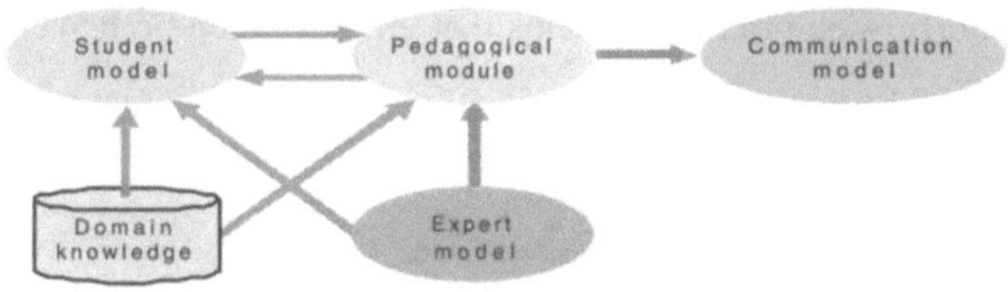

Fig. 1. General concepts of the ITS architecture

ITSs have high intelligent performances. The level of intelligence of an ITS is proportional to the measurement in which the student model describes the real student. The system delivers the educational content to the student based on this model. If the student model contains wrong or incomplete students' profile, the ITS actions would complicate the student learning efforts. Today, this model has to support more sophisticated student properties. These properties are: student interests, educational goals, motivation, social and cultural environment, predisposition, psychological characteristics and many others. If system reactions are based only on the students' results, the system behavior will not be appropriate to the real students' needs. The student model is the ITS metaknowledge about the students (in general). The concrete instances of the student model represent the systems' knowledge about the individual students. An ITS is better if it contains more stereotypes of students' model. Reusability of these entities can be supported by a student ontology.

The cost of high intelligent performance is that many ITSs are strongly focused on one domain. Most of ITSs have a disadvantage that their knowledge base (KB) is only used inside the concrete ITS environment. Therefore these systems do not need a standard representation of their domain knowledge. Usually, a KB is implemented through the rules or constraints. It's also annotated at one kind of script files and which are readable only for specified ITSs. This KB can not be used by other systems. Only ITSs, that support appropriate script format, can reuse this knowledge. Another problem is that knowledge is not described by standard format.

On the other side ILMS inherit the design (building) of learning materials and management abilities from LMSs. While ITSs are concerned about the adaptation to learning possibilities of one student, LMSs are mainly focused on reusability of LOs, and execution of collaborative and administration tasks. ITSs are educational softwares, finalized, and enable the students to improve their skills and knowledge. If a teacher wants to change the learning contents, (s)he has to use an appropriate authoring tool. LMS s support this scenario.

LMSs provide a complete platform in the areas of logging, assessing, planning, delivering contents, managing records and reporting. They improve the both - the self-paced and the instructor-led learning processes [23]. All these activities are represented to the end user (or organization) as a group of Web services. The architecture of LMS is more complex than in the ITS case (Figure 2). As Web services these systems are more transparent and they have more security mechanisms. LMSs are poorly Web oriented systems that are hosted on both Web and application servers.

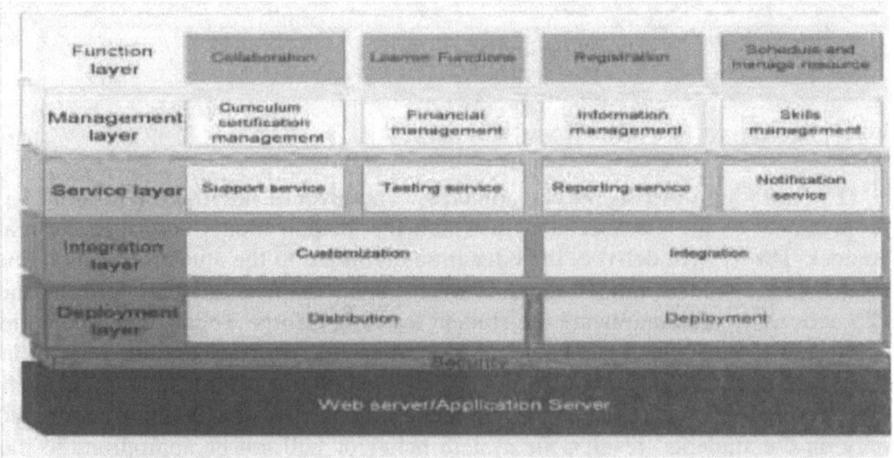

Fig. 2. The LMS Architecture

The last two LMS layers are designed for composing, customizing and communicating services with end-users. This means that LMSs are high-distributed systems over the Web. One course presents an integrated structure of many learning resources that can be hosted on different Web locations. The same resources can be combined with others in different courses. Also, more student groups can learn many courses at the same time. In these conditions, the system must have powerful management features.

This means that an ILMS needs specialized ITS properties and the capacity to perform the described administration, integration and distribution tasks as LMSs. To be more precise, an ILMS has the aggregated structure of the LMS framework, enriched by embedded core of ITS (Figure 3). The ILMS general architecture consists of three basic parts: *administration tools*, *teacher tools*, and *student tools*.

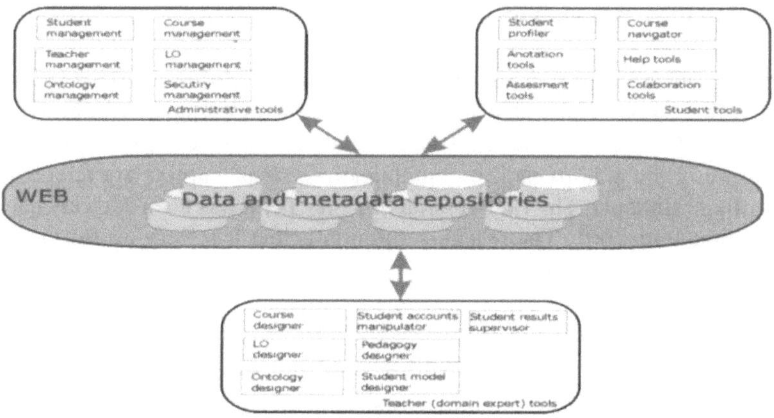

Fig. 3. The ILMS Architecture

The administrative tools support the realization of different management tasks. For example: maintenance of student and teacher records, administration of the domain knowledge and the system security protection.

The teacher tools of the system help teachers to create LOs, combine them with existing LOs and compose the courses. A teacher is responsible for entering students' data and giving the system students' profiles (by creating a specific student model). Domain experts can design the domain ontology that should describe and structure the knowledge (about educational domains, pedagogy and students). The teacher package provides the monitoring of student results that could be used by teachers to track the student sessions.

The student tools generally help students to master the knowledge. The system enables a student to declare his interests, favorites, predispositions and real skills. These data help the system to initiate a student model and determine a student stereotype. While the student uses the system, different tools provide navigation through the learning space, marks for important things, contextual help and skills measurement. The student can also collaborate with other students, teachers and experts.

This is a way that an ILMS provides high cohesion and synergy of efforts from all the subjects in the learning process. The system knowledge is transparent and distributed on the Web. It becomes possible to use concepts of the Semantic Web:

the integration process and the adaptive composing of learning materials. Different specialized pedagogical knowledge becomes accessible for all interested systems over the Semantic Web.

Note also that current LMSs like Blackboard, CourseInfo or WebCT couldn't produce easily the intelligent educational systems, because of the lack of ontological support [10]. They also have abssence of the learner modeling, reasoning and adaptible behavior. Although they provide presentation and management of learning material and scenarios, as well as database management and administration.

7.3 Multitutor: An ILMS

In this section we are trying to present an ILMS named Multitutor. This system is a product of three years research efforts. We started with a single user application, so called *Code Tutor* [34]. This is a small Web-based tutor designed for fast students' briefing in the area of radio-communications. Our learners are telecommunication college students. The first version of Code Tutor has been actively used in classroom since mid-2001. The teachers' opinion is that it is very useful tool, and the students favor this kind of learning.

7.3.1 Motivation

These facts have motivated us to build a new version, which will provide students to communicate with the system through standard Web browsers. The entire system is implemented in Java, using many different current technologies: the CLIPS tool was used for building ES knowledge base files, i.e. Code Tutor's domain knowledge (http://www.ghg.net/clips/CLIPS.HTML), Java-based ES shell *Jess* was used to interpret these files (http://herzberg.ca.sandia.gov/jess/), *Java™ Servlet* technology (http://www.sun.com/products/servlet/) to implement the system's interactions with the students, *Apache HTTP* server (http://www.apache.org/) to store static HTML pages, *Apache JServ* (http://java.apache.org/dist/) to interpret the servlets, and *XML* technology [15] to generate course description files that Code Tutor uses to provide recommendations to the students. Code Tutor is actually Web-enabled and Web-ready, intended primarily for use in the classroom, rather than a full-fledged Web-based ITS built to be used adaptively over the Web.

In the next development phase, we were focused on the authoring tool design. One of the main ITS disadvantages is their narrow domain specialization of the system. For example, the Cognitive tutor [33], which is recommended by NCTM[1], is focused on mathematics (algebra 1 and 2, geometry). The ELM ART [4] system is designed to teach students in LISP programming. The SQL Tutor [28] provides students the possibility to learn SQL. The ILESA [26] system is specialised to teach the solving of linear programming problems. The VALIENT [20] system provides the learning of the data base design.

[1] NCTM – U.S. *National Councel of Teaching Mathematic*

Our opinion is that we developed a domain independent system that provides a useful environment for many courses. This way, we avoided the disadvantage of a rare use of the system. Our goal is to attract many teachers to use Multitutor. Therefore we expect a faster development of this system.

The domain independence is possible only if it is supported with appropriate authoring tools. Above, enumerated ITSs don't provide the teachers the possibility to modify the learning content. On the other side there are a number of authoring tools for ITSs. These are divided in two general groups: *teachers oriented* and *domain oriented*. The first kind of tools provides an easy way for teachers to create courses. The latter type offers a rich interface for describing and structuring the domain. A good example of teacher oriented authoring tool is REEDEM [27]. This tool represents an author-friendly environment, in which, a teacher can define a student model, learning strategies and to describe the course materials. The domain-oriented tools have many possibilities for semantic description of the knowledge. EON [30] is an example of the authoring tool that provides semantic network design of domain concepts and facts (Figure 4). The graphical representation of the resulting domain ontology is very useful, but it demands the teacher to know how to use graphical designer environment.

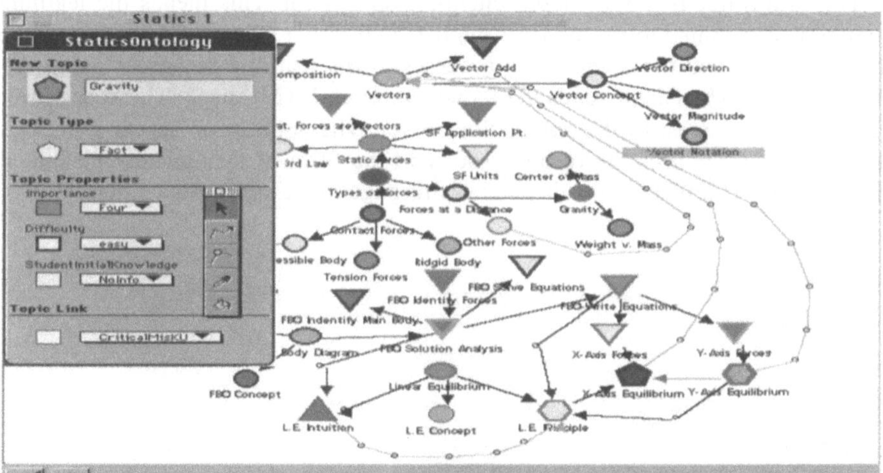

Fig. 4. Design of semantic network in the EON authoring tool

Unfortunately, the created courses can be used only in the specialized framework which is distributed with this tool. The interoperability with the outside applications is nearly impossible because this framework usually does not have interfaces for Dynamic data exchange (DDE) and does not support the standardized data structure formats. The data formats are specialized and strongly coupled with the system components. These frameworks have their own graphical interface and can be hosted only over the LAN environments.

7.3.2 Multitutor Architecture

We tried to design an authoring tool that is a part of the Mulitutor system. The component called *Course Designer* (Figure 7) is designed for this purpose. This tool is accessible to the teachers that want to create their course. We also attempted to formalize the course ontology by using standard describing and structuring format. Our selection is XML as a well-structured format for wide area purposes. The Multitutor system would be sorted in teacher-oriented tools. It provides a course creation without implementation details and course design using appropriate wizards.

The Multitutor is a Web-based client-server system. This means the learning content is distributed to the students via the Web server (Figure 5). The user is on the client side and the student can accesses to the learning resources using the Web browser. The Client sends the request through HTML page. The Web server forwards this request to the application server. The application server processes the request and returns the results usually in the form of dynamically generated HTML page. The Web server dispatches this page to the appropriate client.

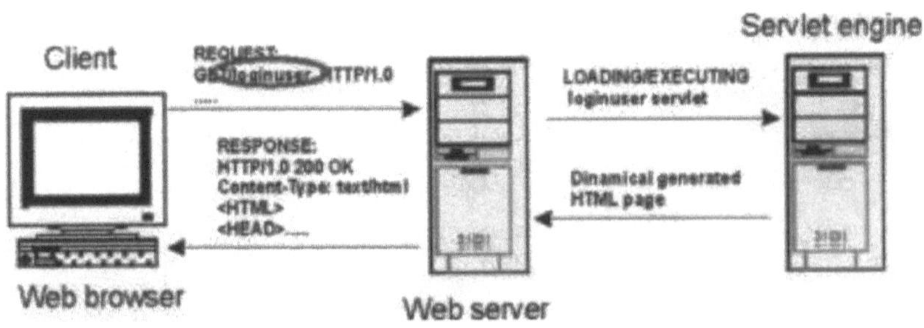

Fig. 5. Client-server paradigm of the Multitutor

The system architecture can be divided in more than three layers (Figure 6). The client's browser can open the HTML pages on any Web location in one session with the system. The Web server proceedes an HTTP request to the application server. Data that are used by the Web applications may be hosted on the different Web repositories. Different applications can use the same data on the Web. One can see in Figure 6 that all the three applications use the domain and pedagogical knowledge from the same network places. This utillity is provided by semantical linked data.

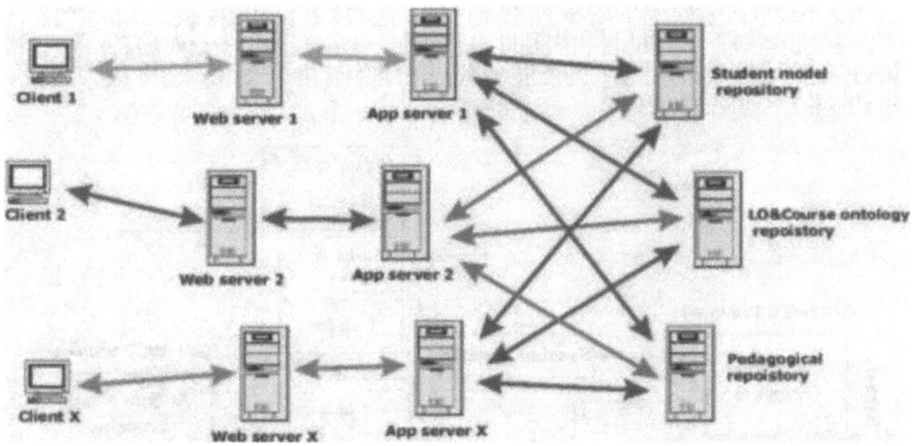

Fig. 6. Part of the real network

Web applications contain the reference maps of Web resources and standardized parsers that can recognize the structure and semantic of these resources. The Web application processes (puts) the user data in correlation with ontologies data from the repositories. This processing can be rule-based (i.e. pattern matching), or based on non-linear reasoning (i.e. fuzzy logic), or other. For example, the applications try to find an adequate stereotype of the current user. Then the application accesses the student model repository and compares the student profile with the accessible models. Based on the founded model, application consults the pedagogical repository to determine appropriate teaching strategy. Then the application can compose learning material for a specified student. The LO and course repository is used for this purpose.

The students can access any Web portal where they have an account. There are three actors in the use-cases of Multitutor: *administrator, teacher, and student.* The administrator executes management tasks in the system. He/she is responsible for:

- Adding and removing the teachers in the tutor system registry – only the registered teachers can use the system;
- Checking the data integrity – controls the teachers and students' registers and log files;
- Viewing the system log files and preserving the system from the malicious user operations – the student results are read-only data and only the course teacher can access and view their results;
- Maintaining the web server and the servlet engine – supervising the Web server repositories and the servlet properties; editing the configuration files and the zone files;
- Maintaining backup of the system files (teachers, students' results, ontology and knowledge base files) – the temporary updating of the copies of the tutor system files.

The teacher tasks are well known. The teacher can create his own courses. These courses can be about different domains. Like as in the LMS, every moment the teacher can monitor his students' results. He can modify the learning contents during the students learning.

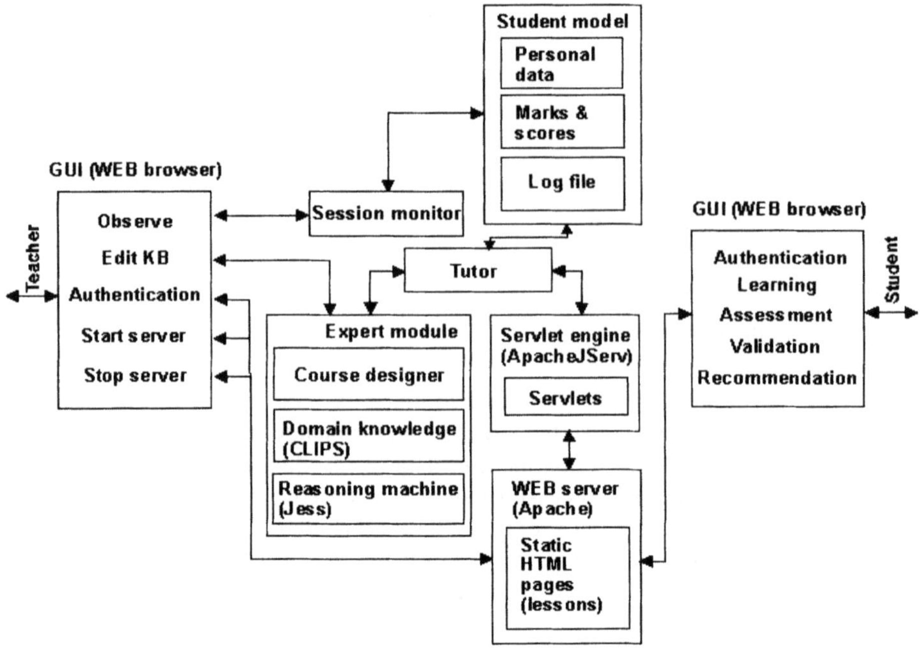

Fig. 7. The Multitutor architecture

The students are organized into a groups (classes) and they access to the courses accordingly to their group. Their communication with the system (logging the system, customizing the interface, learning the course chapters, solving the tests and accepting the skills level and recommendations) runs over the Web browser. The system is designed to support changeable navigation possibilities to the student. It provides the dynamic creation of the learning materials.

The servlet engine represents the application server. The servlets (java classes) play the role of the front end of application. They can refer the functional calls to the middle layer classes. As shown on the model, the core of the system is the tutor concept. The tutor is the main part of the system architecture. It's the system coordinator, dispatcher and monitor at the same time. The pedagogical strategies are implemented in the tutor. It analyzes the data of the student model (model of particular student) and uses its teacher knowledge to require the proper learning contents. The expert module maintains the references of domain knowledge and rule base. The reasoning machine processes the request of the tutor and composes the learning content. This content can include the text, the picture or some other multimedia. In the test phase the content is represented by the test sets or by the problems that students have to solve. These contents the tutor sends back to the servlets. A servlet functional call is broadly propagated over the system. These results with many actions. This way the Multitutor is able to dynamically generate the learning content.

The architecture of the system is designed to support low coupling of particular components. Therefore the system components can be highly distributed over the Web. The administration tasks are performed on the teacher side of application. The administrator interface is not shown because we want to avoid confusion on the diagram. Note that the components are implemented in different technologies.

7.4 Implementation – Multitutor

Based on the low coupling components of the system architecture, the entities are grouped (like a packages) by the functions and data contentment. This section tries to explain the distribution of the metadata.

7.4.1 The Initial System Data

When the system is in use, the tutor module creates a separate instance for every logged student and updates them during the student sessions. The Web server is responsible for delivering the learning contents to a particular student. The initial data that Multitutor uses during the starting phase are stored in the same place (in one file). This file contains the data about the teachers, courses and student groups.

These data provide two things: one is about the registered users (teachers and students) that can use the system, and the other is the path to the course ontology. The initial data are structured to relate teachers, classes (student groups) and courses. The conceptual model (Figure 8) abstracts these relations and it can be translated in the basic system ontology [8].

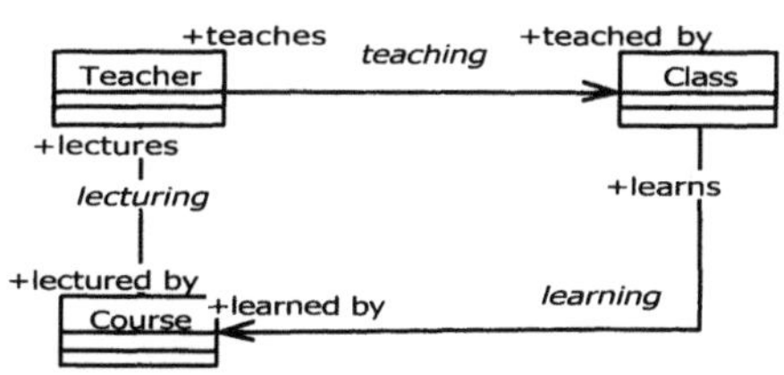

Fig. 8. The general concepts of the learning process

The teacher concept is used in the teacher application. There are two cases: when the teacher creates the course, or when he searches the students' results. The students results is read only.

This model can be converted in an ontology schema that is readable for another part of the application logic. We used XML Schema [15] to create the ontology vocabulary. An excerpt of this XML Schema definition is shown in Figure 9. All the elements are globally defined in the XML Schema definition document. A relation between classes is not defined as an attribute of a class, but as an independent entity, which have a certain domain and range.

```
<?xml version="1.0"?>
<xsd:schema xmlns:xsd="http://www.w3.org/2001/XMLSchema">
    <!-- ... -->
    <xsd:element name="ontotutor">
        <xsd:complexType>
            <xsd:sequence>
                <xsd:element name="Teacher">
                    <xsd:complexType>
                        <xsd:sequence>
                            <xsd:element ref="Name"/>
                            <xsd:element ref="Teaches" maxOccurs="unbounded"/>
                        </xsd:sequence>
                    </xsd:complexType>
                </xsd:element>
                <!-- ... -->
            </xsd:sequence>
        </xsd:complexType>
    </xsd:element>
    <xsd:element name="Teaches">
        <xsd:complexType>
            <xsd:sequence>
                <xsd:element ref="Course"/>
                <xsd:element ref="Class" maxOccurs="unbounded"/>
            </xsd:sequence>
        </xsd:complexType>
    </xsd:element>
    <xsd:element name="Course">
        <xsd:complexType>
            <xsd:sequence>
                <xsd:element ref="Name"/>
                <xsd:element ref="Reference"/>
                <!-- ... -->
            </xsd:sequence>
        </xsd:complexType>
    </xsd:element>
    <xsd:element name="Class">
        <xsd:complexType>
            <xsd:sequence>
                <xsd:element ref="Name"/>
                <xsd:element ref="Student" maxOccurs="unbounded"/>
            </xsd:sequence>
        </xsd:complexType>
    </xsd:element>
</xsd:schema>
```

Fig. 9. Schema of initial application data

Based on the defined metadata in the schema, the Multitutor reads the initial application data from the *ini* file. In the code fragment in Figure 10 the teacher is “Peter Fox” who teaches the course called “Physics” to two classes – “SIG22” and “EW43”. The shown example demonstrates one of the possible ways to represent the relation between the ontological concepts. Many formal and standardized markup languages suffer of the impossibility to represent the semantic of the system data. The same way all other relations from the conceptual model (i.e. from Figure 8) are described by metadata in the schema file. Note that the *Course* concept has an element called *Reference*. This means the data of the specified course are located in one Web destination. Furthermore, the course data are well structured and described by schema (metadata). The *Class* concept semantically represents the student group. The *Student* concept is defined in the student model ontology that is located on a separate repository.

```
<?xml version="1.0"?>
<Ontotutor>
    <Teacher>
        <Name>Peter Fox</Name>
        <Teaches>
            <Course>
                <Name>Physics</Name>
                <Reference>http://servername/mtutor/ontologies/phisics.xml</Reference>
            </Course>
            <Class>SIG22</Class>
            <Class>SIG43</Class>
            <!-- ... -->
        </Teaches>
    </Teacher>
    <!-- ... -->
</Ontotutor>
```

Fig. 10. The initial system data

7.4.2 The Basic Concepts of the Course Ontology

The course has ontology that is referenced in the application *ini* file. The course is an aggregated structure that contains the learning material, the references and the content for assessment. The learning material is structured on the learning objects, which are named *chapters* and *lessons*. Every course is divided on the chapters. Every chapter is divided on the lessons. The lesson is the basic learning unit. One lesson is related to one LO. The learning object is an aggregated structure that consists of the following classes: domain *concept*, *explanation* of the concept, the *learning content* and the *test set* (Figure 11).

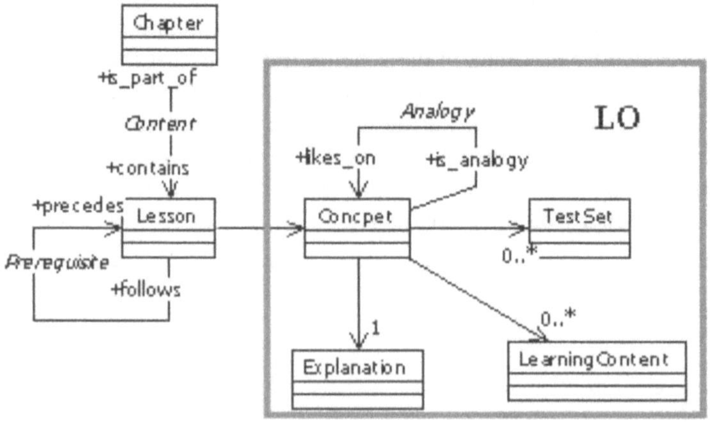

Fig. 11. The main concepts of the course ontology

By this, one LO can be used to create many lessons in the different courses. The LO describes one concept of domain. The concept is related to the explanation, one or more test sets and to the learning contents. The *LearningContent* class represents the multimedia content of the learning object. Depending on different students' knowledge levels the different content will be presented to the student. The concept is self-related. This means one concept is the analogy of some other. The lesson is self-related too. One lesson is the prerequisite to the some other.

The test set is the collection of the questions and related answers that system uses to assess the students' knowledge about one concept. The test set has attributes as ID, level and the test type. The Multitut?r offers the answers to the student. The answers have the marks or the true/false statement. This means the level has to be precisely defined by the course creator (teacher). One LO on the specified level can have number of questions. This way the student gets different questions every time when he repeats the test.

Figure 12 shows a part of the course schema file that is derived from class diagram shown in Figure 11. In the schema we use *Analogies* elements to represent the concept's self-relation.

```
<?xml version="1.0"?>
<xsd:schema xmlns:xsd="http://www.w3.org/2001/XMLSchema">
    <!-- ...-->
    <xsd:element name="ontocourse">
        <xsd:complexType>
            <xsd:sequence>
                <xsd:element ref="Concept"/>
                <xsd:element name="Analogies">
                    <xsd:complexType>
                        <xsd:sequence>
                            <xsd:element ref=" Concept " minOccurs="1" maxOccurs="1"/>
                            <!-- ...-->
                        </xsd:sequence>
                    </xsd:complexType>
                </xsd:element>
                <xsd:element name="LearnigContent">
                    <xsd:complexType>
                        <xsd:sequence>
                            <xsd:element ref="Name" minOccurs="1" maxOccurs="1"/>
                            <!--... -->
                        </xsd:sequence>
                    </xsd:complexType>
                </xsd:element>
            </xsd:sequence>
        </xsd:complexType>
    </xsd:element>
    <xsd:element name="Concept">
        <xsd:complexType>
            <xsd:sequence>
                <xsd:element ref="Name" minOccurs="1" maxOccurs="1"/>
                <!-- ...-->
            </xsd:sequence>
        </xsd:complexType>
    </xsd:element>
</xsd:schema>
```

Fig. 12. The fragment of the course schema file

The entities that are self-related can play different roles. In the next example (Figure 13), there are two lessons in the course *Physics* file (the chapters of the course are not shown). Before the student learns the lesson about the sound waves, he has to learn the lesson about the wave motions.

The analogy is similarly to prerequisite. This self-relation can be used when the student can not pass the tests about the main concept. Then the system tries to explain this concept by the similar one. If the student can not understand the concept of sound waves, the Multitutor helps him by the similar explanation about the water wave. The main goal of analogy is to explain the main concept on the other interesting way. The strong recommendation to the teachers is to use the simpler concepts for the analogies.

```
<?xml version="1.0"?>
<Ontocourse>
    <Course>
        <Name>Physics</Name>
        <!-- ...-->
        <Lesson>
            <Name>Sound Wave</Name>
            <Prerequisites>
                <Lesson>
                    <Name>Wave motion</Name>
                    <!-- ...-->
                </Lesson>
            </Prerequisites>
            <!-- ...-->
            <Concept>
                <Name>Sound Wave </Name>
                <Analogies>
                    <Concept>
                        <Name>Water Wave</Name>
                    </Concept>
                </Analogies>
                <!-- ...-->
            </Concept>
        </Lesson>
    </Course>
    <!-- ...-->
</Ontocourse>
```

Fig. 13. The fragment of the course data

7.4.3 The Student Model

The student model has a separate ontology (Figure 14). This structure has four parts: *the basic student data*, the student stereotype, *students' real skills* (based on the scores) and *the skills that are estimated by the system.* One student can have different skills because he studies many courses. The stereotype holds the sophisticated data about students' interests, favorites, interface customization, the rate of progression, the learning paths, but also data about the most frequently faults. The stereotype is very important for the determining of pedagogic strategy (in the pedagogic module).

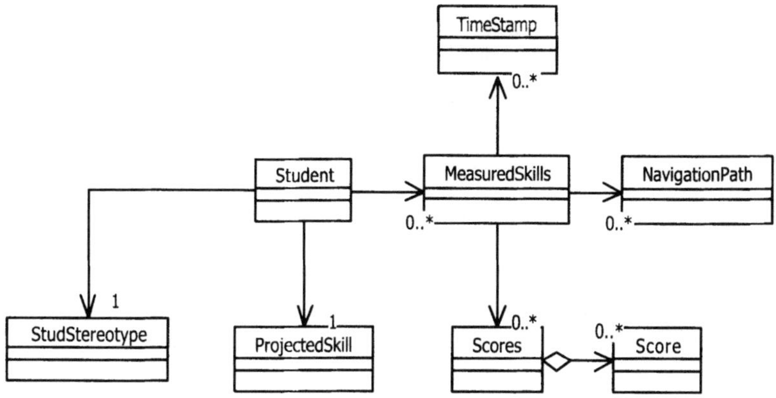

Fig. 14. The student ontology

The relations are uniformly propagated through the model in the student ontology. Multitutor sorts a student in one stereotype. The student skills are determined when the student starts to use the system. During the first session the student gets the questionnaire and the pretest. Those results are used to predict the student success and they are represented by the *ProjectedSkill* concept of the model (Figure 15). While the student learns the course the system monitors the students' navigation and time which is spent on the studying every particular concept. The student gets the tests and Multitutor serializes the results. The *MeasuredSkill* concept provides the correlation of the students' data. Those data are processed by the expert module and the conclusions are used by the pedagogical module to compose the next learning content.

```
<?xml version="1.0"?>
<xsd:schema xmlns:xsd="http://www.w3.org/2001/XMLSchema">
    <xsd:element name="ontostudent">
        <!-- ... -->
        <xsd:complexType name="Skill" abstract="true">
            <!-- ... -->
        </xsd:complexType>
        <xsd:complexType name="ProjectedSkill">
            <!-- ... -->
            <xsd:extension base="Skill">
        </xsd:complexType>
        <xsd:complexType name="MeasuredSkill">
            <!-- ... -->
            <xsd:extension base="Skill">
        </xsd:complexType>
        <xsd:complexType>
            <xsd:sequence>
                <xsd:element name="Student">
                    <!-- ... -->
                </xsd:element>
                <xsd:element name="PreferedLearningStyle" type=" ProjectedSkill ">
                    <!-- ... -->
                </xsd:element>
                <xsd:element name="Test" type=" MeasuredSkill ">
                    <!-- ... -->
                </xsd:element>
                <!-- ... -->
            </xsd:sequence>
        </xsd:complexType>
    </xsd:element>
    <!-- ... -->
</xsd:schema>
```

Fig. 15. The student model metadata

7.4.4 The Learning Content

The learning materials are dynamically composed (Figure 16). The Multitutor correlates the current and historical student data and it makes decision about the learning content. The basic explanation of the domain (lesson) concept is in the text form (The *Explanation* entity in Figure 11).

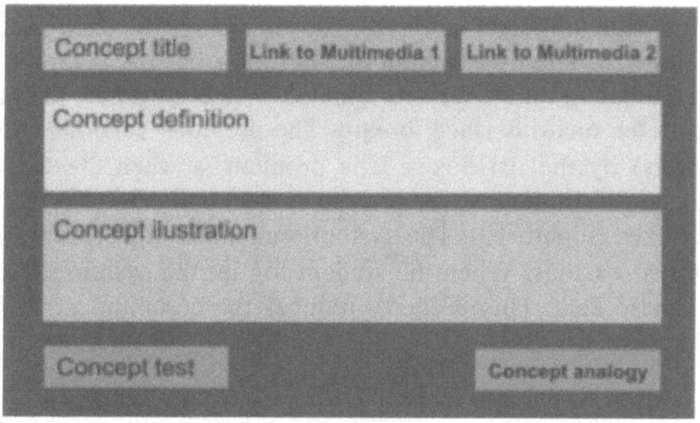

Fig. 16. The Web page structure

The text is yet the most precisely way to define the concept and avoid the ambiguity. In the Multitutor, the other contents (figures, sounds, and video) are used to support the better understanding and the faster learning of the concepts. The learning material is represented by Web page that has the table structure.

7.5 Mutitutor Applications

This chapter presents the description of the three courses which are designed by using the Multitutor.

7.5.1 Code Tutor

The Multitutor is used to compose three different courses. The first of them is Code tutor that is designed for learning the protocols and codes that are used in the radio-communications. The student starts the session by log in the system (Figure 17). He selects the class and types his name and password.

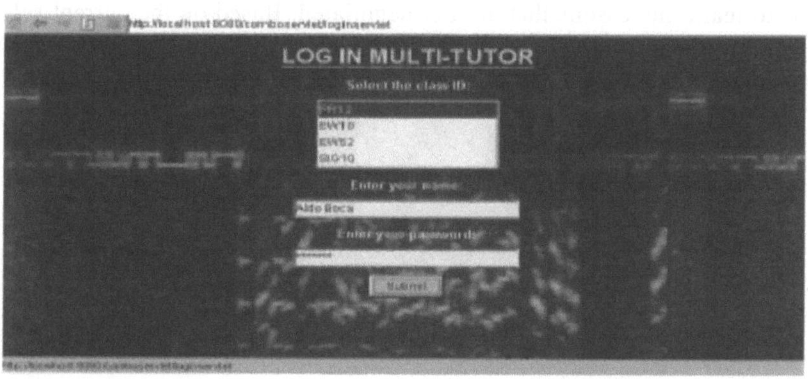

Fig. 17. The student login

The student adapts the page lookup (Figure 18). He can change the background texture, the font style and size. The first version of Multitutor has the yellow font and the blue texture in the background. We are founded by the inquiries, the half of the students do not prefer such interface lookup. The previous versions show the test (possible answers) by the list boxes. The problem is when the text is greater than the list box width. In the last version the student can select the radio-button style or the list boxes (Figure 19). The system remembers the students' selections from the previous sessions. When the student log in, the system presets the interface based on these data. This property reduces the communication between the students and system.

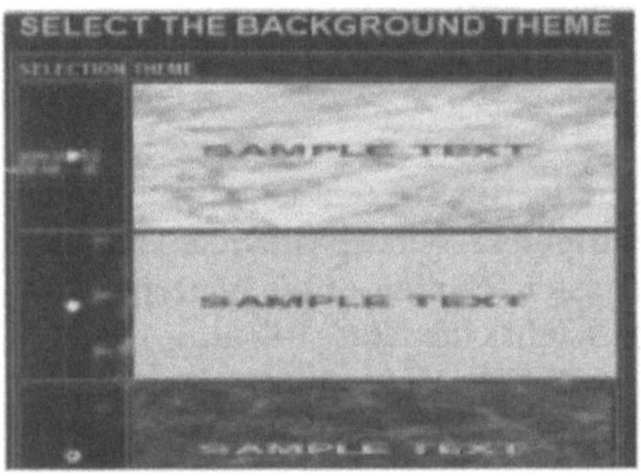

Fig. 18. The background and text setup

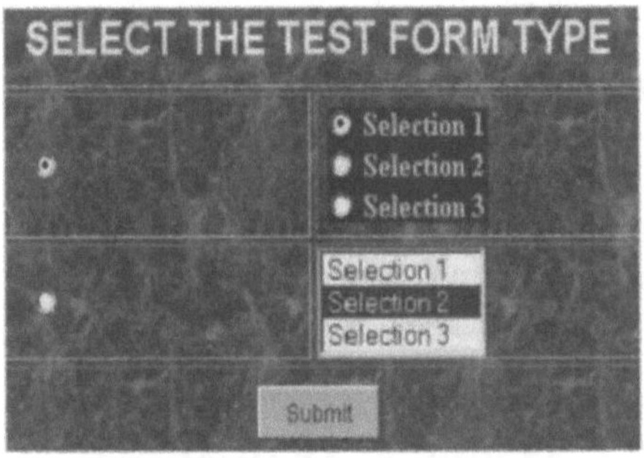

Fig. 19. The test form styles

The Multitutor offers the courses that are designed for the students' class. In the next step, the student selects the course. In the one session the student can learn one course. If the student wants to learn another course, he has to finish the session and starts the new one.

After the student selects the course, he has to choose the chapter. In the one session the student can learn more chapters of the one course. He can also learn the same chapter more times.

The next stage is the learning (Figure 20). The chapter is divided on to the lessons. The sequence of the lessons is defined from the student model data. The student has to learn the lessons that are not negotiated. Based on the current selections and the data from the previous sessions, the Multitutor composes and delivers the learning content to the student.

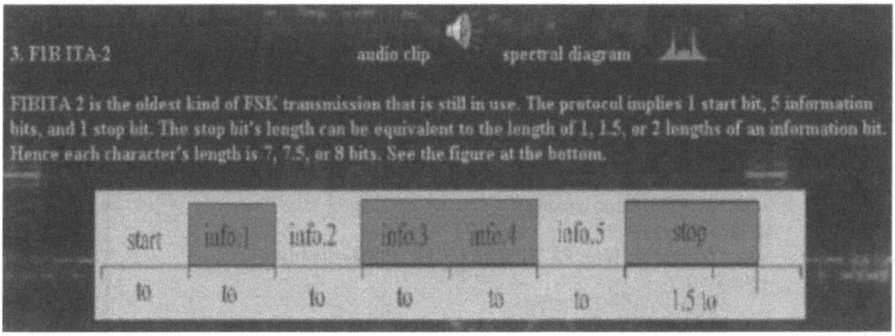

Fig. 20. The learning content

The learning material consists of five basic parts: The title of the concept, the basic explanation, the main illustration that describes the concept, the other multimedia content (e.g. audio clips and figures), and the links. There are two links: the analogy and the test link.

The student can learn the same lesson arbitrary times. The next phase is the testing (Figure 21). The Multitutor delivers the test questions to the user. The test content is related to the one-chapter lessons. If the student learns the same chapter more than one, every time he gets the new set of questions. There are tree types of the test sets in the system: the multiple selections, the single selection and the answers scaled by marks. In the case of the multiple/single selection types the system calculates the percentage of efficiency. In the third case, the system calculates the average of the particular marks. In the next stage, the system represents the results to the student. If the student is failed he has to repeat the bad marked lessons again. If the student passes the test, the system recommends him what to learn. In this case the student can accept this or select another chapter.

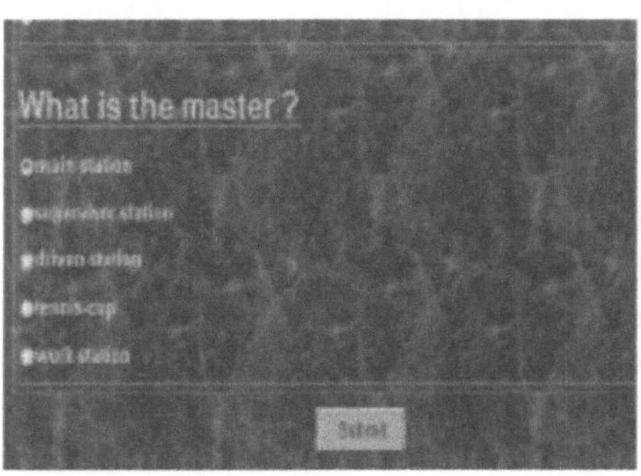

Fig. 21. The test page

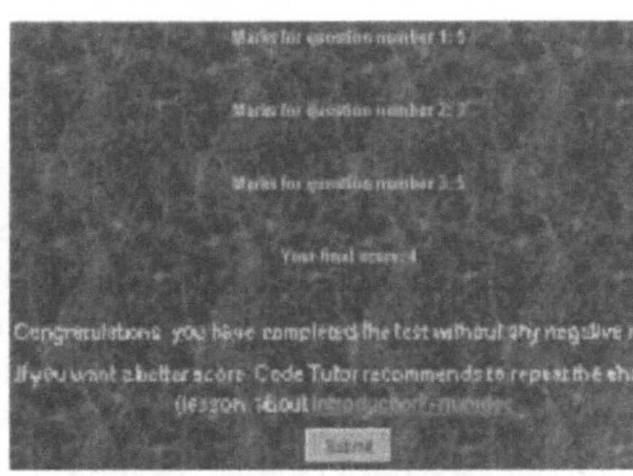

Fig. 22. The results and recommendation

Before the system represents the results to the student (Figure 22), it saves these data in the XML-formatted result file (Figure 23). In the previous versions of Multitutor, the results are grouped by the course. To support the better student model, the new version creates the results file for each student. The student can see the test details by clicked on the question result. Than the system opens the new page that contains the question, student answer, correct answer and mark.

```
<?xml version="1.0"?>
<results>
    <student>
        <name>Aldo_Boca</name>
        <class>EW52</class>
        <testdate>Mon_Mar_22_12:28:01_CET_2004</testdate>
        <tcourse>
            <name>code-3</name>
            <tchapter>
                <name>decoding</name>
                <tlesson>
                    <name>artrac</name>
                    <setnum>1</setnum>
                    <level>1</level>
                    <tquestion>What time (ms) of artrac CRC is ?</tquestion>
                    <tanswer>1016</tanswer>
                    <tbestanswer>706</tbestanswer>
                    <tmark>2</tmark>
                </tlesson>
            </tchapter>
        </tcourse>
        <!-- ... -->
    </student>
    <!-- ... -->
</results>
```

Fig. 23. The student results

The Multitutor provides XSLT (Extensible Style Language Transformation) to convert results from XML format to HTML format and represents them through the Web page. A student can see only his/her results, but the teacher can see the results of the whole class (classes). These data are shown in the read only format both the teacher and the student.

7.5.2 Design Pattern Tutor

Design Pattern Tutor is a Multitutor course for learning Design Patterns. The course gradually introduces student with the concept of design patterns and describes most frequently used classes of patterns.

The issue imposed in course realization for learning the Design Patterns, is the way of organization of learning process. In the course of realization of this problem, the "Design Patterns" book [16] should support this course. The course for learning Design Patterns is divided into three main sections: creation patterns, structural patterns and behavioral patterns.

The ITS system for learning the Design Patterns, described in this section, makes tutorial model of learning the Design Patterns possible, as well as independent study of patterns for more advanced students. The system provides an intelligent representation of educational material. This material is adjusted to student performance, such as degree of backward knowledge, desirable detail level and assessments of the system. Also this system assess the level of student's acquaintance with this domain.

During interaction of a user with the system, the Tutor monitors the performance of the student. It updates the Student model and checks if the plan is still appropriate. The Tutor uses tests and exercises to get feedback from the student in order to infer changes in the Student model. Assessments obtained through such interactions are compared with the other parameters and assessments in order to get the final judgment of a student.

If a student is having a session with Tutor for the first time, system gives him a set of exercises and tests before the tutoring process starts. In this way, the student model estimates the new student's background knowledge and determines initial values. After the initial assessment of the student, knowledge system runs in teaching mode. At the beginning of this mode, system develops an initial plan of teaching actions. The plan is based on the contents of the lesson being presented, the relevant pedagogical rules and the student model. The plan represents a detailed outline of the lesson presentation.

At the end of presented topic, the Tutor runs at examination mode. It generates exercises and tests for the student and assesses his knowledge. During the session, system observes and adapts the student progress with the generated course. If the student answers the test items correctly, he progresses along the course and no changes to the course are necessary. However, if the student fails to answer the test items correctly, the system must modify student performances. If the student's performance does not meet expectations, the course is dynamically re-planned. Through dynamic regeneration, each student is able to get a highly personalized course for his needs. [7].

The system must memorize each activity of a user and the system, as well as all assessments of a student, updating the student model. These data may be used to prepare instructional plan, as well as to give advice and recommendations for further work [32].

During the session, system observes and adapts the student progress with the generated course. Domain model of Design Pattern Tutor is made up of concepts, which correspond to one pattern. Each concept is divided in units – the elementary pieces of domain knowledge. There are a fixed number of units in particular concept, but the size of unit is not fixed. The system uses unit variants technique, which consists of keeping two or more alternative pages with adapted content, e.g. one for each knowledge level: beginner, intermediate and expert. Each unit has an arbitrary number of fragments – a chunk of information that should be presented to the user. The user model and the concept relationships of the domain model provide the information that allows the system to determine which chunk of information should be presented to the user. The chunk of information may also consists of fragment variants, i.e. fragments related by an "or" relationship.

The system provides students with two kinds of navigation through the course material (Figure 24.):

- Direct guidance – The student sees only one options to continue with the browsing activity i.e. just one button to navigate to the "next" page is displayed. The destination of the "best" link is determined by the system.

- Link removing – advanced students could choose which topics to learn by selecting appropriate link from content menu, but links that the system considers inappropriate are removed, i.e. they are not longer available. Anchors of these links are replaced by text.

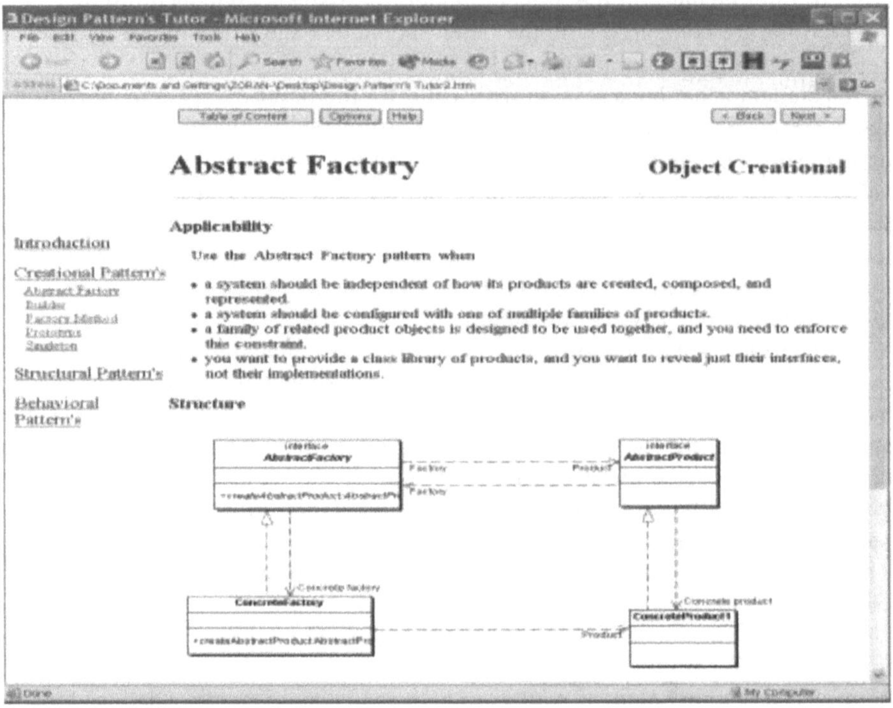

Fig. 24. A Web page shows a screenshot of Design Pattern Tutor - course created by Multitutor

The student has the option of letting the Tutor choose the next topic or choosing it himself. In both cases, the student must achieve sufficiently ready score for the topic. The topics are represented in a dependency graph, with links representing the relationship between topics, which include prerequisite and related topic. A student is ready to learn a topic only if he has performed sufficiently well on its prerequisites.

At the end of each topic student has to complete test. If the student fails to give correct answers, the Tutor must provide an alternative learning path to the student, such as a hint. If the learner tries to move on to a new topic before the Tutor founds that the student has explored the current topic sufficiently. The Tutor will generate a warning, suggesting better exploration of the current topic. These warnings also remind the student of the availability of hints. The student can choose either to follow the advice or to move on.

7.5.3 Semantic Web Empowered Learning: A support for Petri Nets

In order to illustrate how Multitutor can be used for Semantic Web learning applications we show a simple Petri net educational system. The Petri net system is intended to be used in a number of computer science courses that use Petri nets (e.g. distributed computer systems, computer architecture, operating systems, etc.). However, if we want to use Petri net model in Multitutor we should prepare suitable equipment. In our case we have the following elements:

- *The Petri net ontology* – This ontology we assume as a domain ontology of Petri net educational context. This ontology is developed using Protégé tool and UML. The ontology describes Petri net conceptualization using RDFS and OWL languages. Additionally, the Petri net ontology is in accordance with the Petri Net Markup Language (PNML) – an ongoing Petri net community effort for the standard XML-based sharing format. The Petri net ontology has a common part that contains concepts common for all Petri net dialects. Afterward, this common part should be specialized for concrete Petri net dialect [17].
- *The P3 tool* – A Petri net tool we have developed for teaching Petri nets [18]. The P3 tool supports the two Petri net dialects: P/T nets and Upgraded Petri nets. Also, it has the following Petri net analysis tools: reachability tree, equation matrix, firing graph, firing tree. The P3 tool has advanced model-sharing features based on PNML. Furthermore, it has a collection of the XSLTs that transform PNML to other Petri tool specific formats (i.e. DaNAMiCS, Renew, Petri Net Kernel, and PIPE). Also, P3 implements conversion of the PNML Petri net model description to Scalable Vector Graphics (SVG). Since this format can be viewed in standard Web browsers (e.g. Internet Explorer), it is suitable for creating Web-based Petri net teaching materials. Learning objects, created in this way, have their underlying semantics described in RDF form, and we are able to transform them (e.g. using XSLT) into PNML. That way, the learning object can be analyzed with standard Petri net tools.
- *Petri net Web Service* – A Web service that uses a PNML Petri net model as input, performs one simulation step and generates result, again, in PNML format [21].

The resulting Semantic Web infrastructure for Petri nets is shown below (Figure 25). This infrastructure summarizes all major features of the Petri net ontology, P3, and the Petri net Web Service. The central part of this infrastructure is PNML.

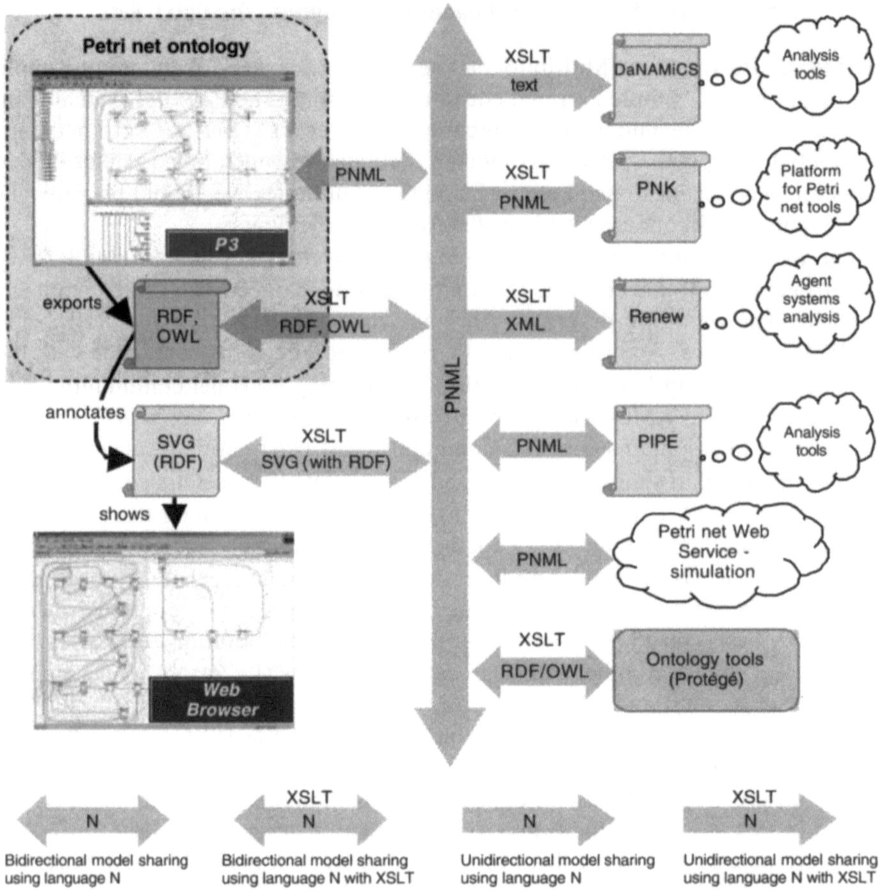

Fig. 25. Petri net infrastructure for the Semantic Web (that uses "PNML-based bus" for model sharing): the Petri net ontology, current Petri net tools, P3 tool, Web-based applications, Petri net Web Service, and ontology tools for validation of Petri net documents using the ontology

This Petri net infrastructure for Semantic Web can have a number of applications in practice. Here, we show how it can be used within Mult itutor for developing Petri net courses. A teacher creates Petri net models in RDF-annotated SVG format using the P3 tool. Then, the teacher uses these models in Multitutor where he/she creates courses following the procedure that we have already explained. In fact, the teacher use Petri net models as figures, but these figures have an ontologically annotated content. After the teacher have finished a course students can use it for studying.

In order to empower Multitutor created courses with ability to perform interactive simulation of Petri net models, implementation of the logic of Petri net execution is needed. This can be achieved using Web service for Petri nets simulation developed. From the created course we should forward a Petri net model to the Web service. This model is converted from RDF annotated SVG format into PNML format using an XSLT. Once the simulation is finished, another XSLT is used to transform the result from PNML to RDF annotated SVG format. Both XSLTs are part of proposed infrastructure. The suggested approach to educational systems development using proposed Petri net infrastructure for the Semantic Web is depicted below (Figure 26).

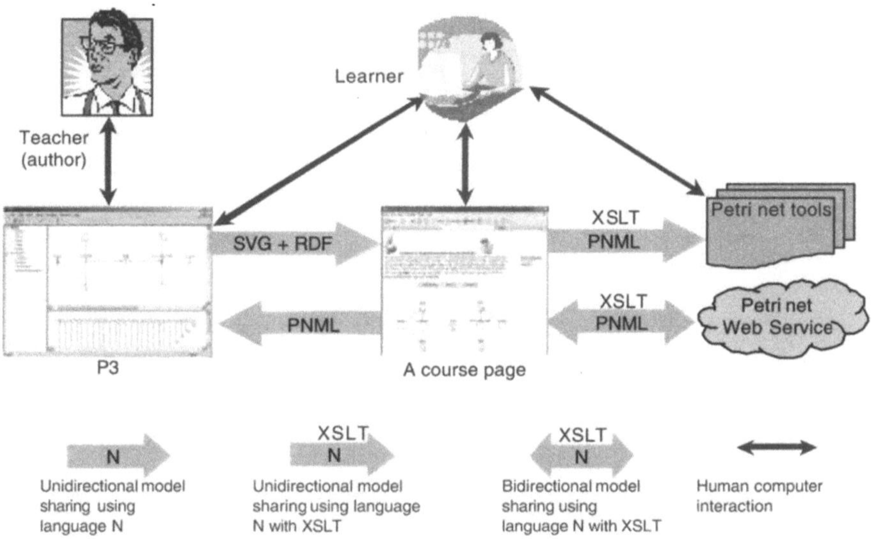

Fig. 26. An approach to using Petri net infrastructure for the Semantic Web in Web-based educational systems

Note that we can not implement the calls of Web service procedures from Multitutor. Actually, we should extend generated Multitutor's courses manually by adding a few method calls responsible for using Web service's methods. In the future Multitutor versions we are planning to implement suitable tools that will be able to access and use Web service.

The Web page from the lesson that helps students to understand and learn the well-known producer/consumer synchronization problem is shown below (Figure 27). This problem is a common part of many different courses in computer science. User begins his/her interaction with the Web page by pressing button **Initial Marking** in order to define initial marking of the Petri net model. Automatically, Petri net graph conforms to the specified data reflecting changes of the model. A press on the **Simulate** button is a sign for system to start simulation of the model. Simulation is performed in collaboration with the Web service according to the previously explained scenario. Simulation results are shown on the Petri net graph. User can save a Petri net he/she is working with in PNML format by choosing button **Save as**. Having studied this Web page the student is going to the next one according to the course plan.

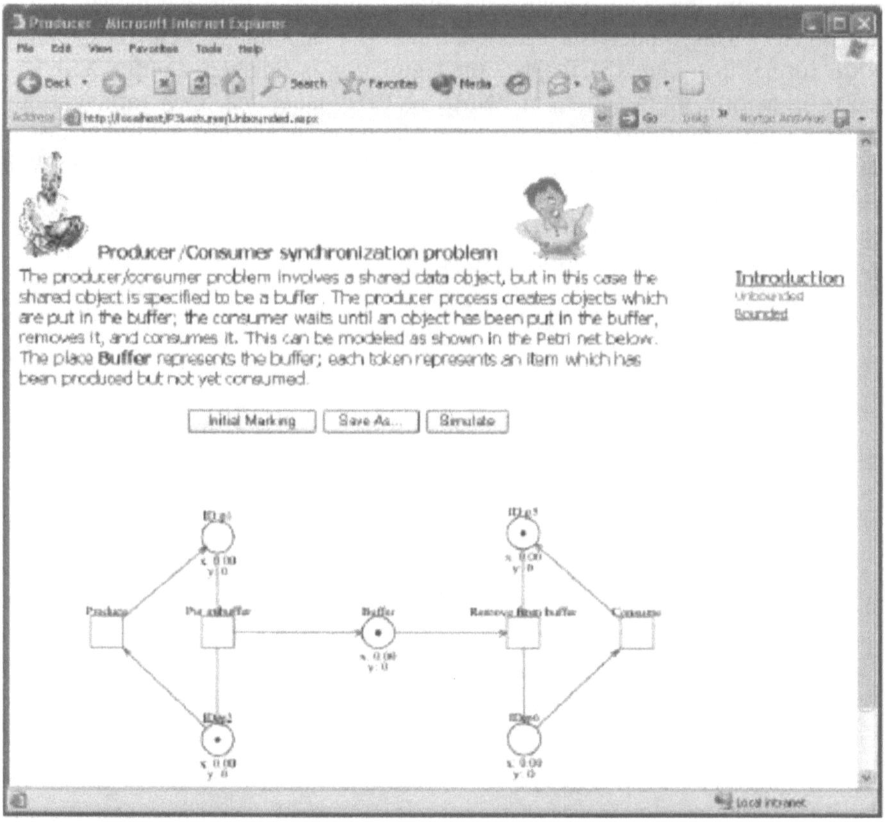

Fig. 27. A Web page shows how RDF-annotated SVG documents can be used in courses created by Multitutor

7.6 Intelligent Learning Management Systems and the Semantic Web: Future Improvements

We have so far shown the main features of the Multitutor system as well as examples of two learning applications developed in the Multitutor. We especially stressed how the Multitutor describes metadata regarding their interoperability. Accordingly, we have explained three XML Schemas that describe: 1. The whole system (Figure 9), 2. Courses, 3. Student models. However, the XML Schema mechanism itself has several weaknesses regarding the ontology description [24], so in the future Multitutor versions we should improve some of them. The main point is to use the Semantic Web ontology languages (e.g. RDF(S) and OWL) as well as e-learning initiatives and proposals based on those languages. Here we shortly elaborate some important experiences that can be useful for the future Multitutor improvements.

Edutella is a democratic (peer-to-peer) network infrastructure for search and retrieval of information about learning resources on the Semantic Web [31]. Brase and Nejdl showed how ontologies could be exploited to enhance LO metadata in Edutella [3]. They gave an example of an ontology developed in accordance with the ACM Computer Classification system (ACM CSS). This ontology was described with RDF, and used in the Edutella system. The ontology improved the searching for leaning objects and it would be a useful for Multitutor. The navigation through learning materials as well as their findabilty can be improved by topics maps [12]. Topic maps provide a language to represent the conceptual knowledge with which a student can distinguish learning resources semantically. Moreover, topic maps are very suitable for representing the course unit ontological structure.

The EU/ITS project ELENA (http://www.elena-project.org/) tries to provide solutions for personalization, openness, and interoperability in the context of smart spaces for learning [13]. This project emphasize that we should use appropriate standards to describe a learner profile. Examples of attempts to standardize a learner profile are IEEE Personal and Private Information (PAPI) (http://ltsc.ieee.org/wg2/) and IMS Learner Information Package (LIP) (http://www.imsproject.org/profiles/index.cfm). Taking into account these two standards the authors' of the Elena project developed the learner ontology. The ontology keeps information about appropriate learning resources which are relevant with respect to user interests, user performance in different courses within one domain or even different domains, user goals and preferences, etc. This ontology in the RDFS form is available at http://www.learninglab.de/~dolog/learnerrdfbindings/. Another useful direction for describing student models in Multitutor as well as on the Semantic Web is the User Modeling Markup Language (UserML) [22]. UserML is an ontology-aware XML vocabulary defined by the UserOL ontology.

Several Educational Modeling Languages (EMLs) have been recently emerged. One of EML definitions states that an EML is a semantic notation (i.e. metamodel or ontology) for units of learning to be used in e-Learning [25]. They have XML binding and they are pedagogically flexible. The final result of an EML should be an instructional model with the following segments: content, didactical (e.g. sequencing) and presentational [36]. These EMLs attempts can be used as guidelines how Multitutor courses can be described in the future. In fact, we can use an EML instead of the Multitutor's course ontology.

Note that the learning technology community lacks standardized-ontologies for all these described aspects. However, all these efforts give useful guidelines for the future improvements. We believe that a solid starting point for new Multitutor versions is to use RDFS defined annotations instead of current XML Schema based formats.

7.7 Conclusions

In this chapter we tried to explore development of ILMSs for the Semantic Web. As result of our research we developed Multitutor an ILMS that uses XML-based technologies (i.e. XML Schema and XSLT) in the combination with the well-proven tools for developing intelligent systems (i.e. Jess). Our first experience with Multitutor is encouraging from both – the student side and the teacher side. However, our ILMS needs further changes in order to better exploit the Semantic Web benefits (e.g. we should use RDFS or OWL definitions of course and student ontologies rather that current XML Schema definitions). Ontology development and Semantic Web languages for e-learning (e.g. Edutella, Elena, UserML, Topic Maps, etc.) can be very useful in this direction. Note that many author in the e-learning community defined ontologies of different kinds of knowledge [29] in the last years. But, this makes the problems for developers like which solution is the most appropriate. Accordingly, the main challenge for the e-learning community is to adopt standard Semantic Web ontologies [11] that will be guidelines for the developers of LMSs/ILMSs.

References

[1] Beck J., Stern M. and Haugsjaa E.,(1996): Applications of AI in Education, ACM Crossroads, Vol. 3, No. 1, pp. 11-15

[2] Berners-Lee T., Hendler J., Lassila O.,(2001): The Semantic Web, Scientific American, Vol. 284, No. 5, pp 34-43.

[3] Brase J., Nejdl W.,(2004) Ontologies and Metadata for eLearning, In S. Staab & R. Studer (Eds.) Handbook on Ontologies, Springer-Verlag, pp. 555-574.

[4] Brusilovsky P., Schwartz E. and Weber G.,(1996): ELM-ART: An Intelligent Tutoring System on the World Wide Web, In Proceedings of the 3rd International Conference on Intelligent Tutoring Systems, Montreal, Canada, pp. 261-269.

[5] Brusilovsky P.,(2001): Adaptive Hypermedia, User Modeling and User-Adapted Interaction, Vol. 11, No.1-2, pp. 87-110.

[6] Brusilovsky P.,(2003): A Distributed Architecture for Adaptive and Intelligent Learning Management Systems, In Proceedings of the AIED 2003 Workshop Towards Intelligent Learning Management Systems, Sydney, pp. 5-13.

[7] Brusilovsky P., Vassileva J.,(2003): Course sequencing techniques for large-scale web-based education, Int. J. Cont Engineering Education and Lifelong Learning, Vol. 13, Nos. 1/2, pp. 75-94.

[8] Calvo R. A.,(2003): User Scenarios for the design and implementation of iLMS, In Proceedings of the AIED 2003 Workshop Towards Intelligent Learning Management Systems, Sydney, pp. 14-22.

[9] Devedžic V.,(2003): "Web Intelligence and AIED," In Proceedings of the AIED 2003 Workshop Towards Intelligent Learning Management Systems, Sydney, pp. 23-33.

[10] Devedžic V.,(2003): Key Issues in Next-Generation Web-Based Education, IEEE Transactions on Systems, Man, and Cybernetics – Part C: Applications and Reviews, Vol. 33, No. 3, pp. 339-349.

[11] Devedžic V.,(2003): Think ahead: evaluation and standardization issues for e-learning applications, International Journal of Continuing Engineering Education and Lifelong Learning, Vol. 13, No. 5/6, pp. 556-566.
[12] Dichev Ch., Dicheva D. D. and Aroyo L.,(2004): Topic Maps for E-Learning, International Journal on Advanced technologies for Learning, ACTA Press, Vol. 1, No. 1, pp. 1-7.
[13] Dolog P., Henze N., Nejdl W. and Sintek M.,(2004): Personalization in Distributed eLearning Environments, In Proceedings of the 13th International World Wide Web Conference, NY, USA.
[14] Duval E., Hodgins W., Sutton S.A. and Weibel S.,(2002): Metadata principles and practicalities, D-Lib Magazine, Vol. 8, No. 4.
[15] Fallside D. C., ed. (2001): XML Schema Part 0: Primer, W3C Recommendation [Online]. Available: http://www.w3.org/TR/2001/REC-xmlschema-0-20010502/
[16] Gamma E., Helm R., Johnson R. and Vlissides J., (1995): Patterns –Elements of Reusable Object-Oriented Software, Addison-Wesley Publishing Company, USA.
[17] Gaševic D. and Devedžic V.,(2004): Reusing Petri Nets Through the Semantic Web, In Proceedings of the 1st European Semantic Web Symposium, Heraklion, Greece, pp. 342-351.
[18] Gaševic D. and Devedžic V.,(2004): Teaching Petri Nets Using P3, Educational Technology & Society (Journal of IEEE Technical Committee on Learning Technologies), (forthcoming).
[19] Gruber T. R.,(1993): A translation approach to portable ontology specifications, Knowledge Acquisition, Vol. 5, No. 2, pp. 199-220.
[20] Hall L. and Gordon A.(1998): Synergy on the Net: Integrating the Web and Intelligent Learning Environments, In Proceedings of The Workshop on Web-Based ITS, San Antonio, TX, pp. 608.
[21] Havram M., Gaševic D. and Damjanovic V.,(2003): A Component-based Approach to Petri Net Web Service Realization with Usage Case Study, In Proceedings of the 10th Workshop Algorithms and Tools for Petri nets, Eichstätt, Germany, pp. 121-130.
[22] Heckmann D., Krueger A.(2003): A User Modeling Markup Language (UserML) for Ubiquitous Computing, In Proceedings of the 9th User Modeling Conference, Johnstown, Pennsylvania, USA, pp. 393-397.
[23] iCMG Learning Management System (LMS) Architecture (May 25, 2004) [Online]. Available: http://www.icmgworld.com/corp/ces/ces.lms.asp
[24] Klein M.(2001): XML, RDF, and Relatives, IEEE Intelligent Systems, Vol. 16, No. 2, March/April, pp 26-28.
[25] Koper R.(2002): Educational Modeling Language: adding instructional design to existing specifications, Workshop "Standardisierung im eLearning", Frankfurt, Germany.
[26] López J. M., Millán E., Pérez-de-la-Cruz J.L. and Triguero F.(1998): Design and Implementation of a Web-based Tutoring Tool for Linear Programming Problems, In Proceedings of the Workshop on Web-Based ITS, San Antonio, TX, [Online] Available: http://www-aml.cs.umass.edu/~stern/webits/itsworkshop/ilesa.ps, pp. 1-8.
[27] Major, N.(1995): REDEEM: Creating Reusable Intelligent Courseware, In Proceedings of The International Conference on Artificial Intelligence in Education, Charlottesville, VA, USA, 1995, pp. 75-82.
[28] Mitrovic A. and Hausler K.(2000): Porting SQL-Tutor to the Web, In Proceedings of the International Workshop on Adaptive and Intelligent Web-based Educational Systems, Montreal, Canada, pp. 50-60.
[29] Mizoguchi R. and Bourdeau J.(2000): Using Ontological Engineering to Overcome Common AI-ED Problems, International Journal of Artificial Intelligence in Education, Vol. 11, pp. 1-12.

[30] Murray T.(1996): Having It All, Maybe: Design Tradeoffs in ITS Authoring Tools, In Proceedings of the 3rd International Conference on Intelligent Tutoring Systems, Montreal, Canada, pp. 93-101.
[31] Nilsson M., Palmér M. and Naeve A.,(2003): The Edutella P2P Network - Supporting Democratic E-learning and Communities of Practice, in McGreal, R. (ed.) Accessible education using learning objects, Taylor & Francis Books Ltd., London, UK, pp.78-85.
[32] Prentzas J., Hatzilygeroudis I., Garofalakis J.(2002): A Web-Based Intelligent Tutoring System Using Hybrid Rules as Its Representational Basis, In Proceedings of the 6thIntern. Conference, ITS France and Spain, pp. 119-128.
[33] Ritter S.(1997): PAT Online: A Model-Tracing Tutor on the World-Wide Web, In Proceedings of the Workshop-a Intelligent Educational Systems on the World Wide Web, Kobe, Japan, pp. 11-17.
[34] Šimic G., Devedžic V.(2003): Building an intelligent system using modern Internet technologies, Expert Systems with Applications, Vol. 25, No. 2, pp. 231–246.
[35] Stojanovic Lj., S. Staab S. and Studer R.(2001): eLearning in the Semantic Web, In Proceedings of the World Conference on the WWW and the Internet (WebNet 2001), Orlando, Florida, USA, pp. 325-334.
[36] Weitl F., Süß C., Kammerl R.and Freitag B.(2002): Presenting Complex e-Learning Content on the Web: A Didactical Reference Model, In Proceedings of World Conference on E-Learning in Corporate, Government, Healthcare, & Higher Education, Montreal, Canada, pp. 1018-1025.

8. Developing a User Centered Model for Creating a Virtual Learning Portfolio

Michael Verhaart[1] and Dr Kinshuk[2]

[1] Faculty of Business and Computing,
Eastern Institute of Technology, Hawke's Bay, New Zealand,
mverhaart@eit.ac.nz

[2] Director Advanced Learning Technologies Research Centre,
Information Systems Department, Massey University, New Zealand,
kinshuk@ieee.org

This chapter presents a personal content management framework, where an individual has the ability to create an electronic summary of their knowledge. This includes adding facts and information from research, plus personal knowledge and insights. If both instructor and learner work within some or all of the components of the framework, this will greatly enhance the ability to share content, both from instructor to learner and learner to learner.

This chapter covers the structure of the personal content management framework, how the "Me" model allows for the management and capture of knowledge by using content objects known as "sniplets", that may have attached Multimedia objects described using MVML. The chapter looks at the evolution of the model, observations and feedback from learners using an existing prototype and possible future directions.

8.1 Introduction

A common way in which our individual learning is managed is through the creation of a personal portfolio of facts, information, knowledge and insights. Traditionally this has been achieved through summarising content, discussion, observations, and/or organising them in a paper based or electronic system. With the huge explosion of content available, particularly on the World Wide Web, new techniques need to be developed to help manage this personal learning content.

This chapter discusses a personal content management framework, where individuals have the ability to create a personal electronic portfolio of their knowledge. This includes adding facts and information from instruction, research, discussion, experience, insights and feedback. If some or all of the components are adopted by instructors, learners and content suppliers, the ability to create and share knowledge will be greatly enhanced. Using web based technologies and its ability to be accessed by both the knowledge owner and authorized external users,

M. Verhaart and Dr. Kinshuk: *Developing a User Centered Model for Creating a Virtual Learning Portfolio*, StudFuzz **178**, 203–232 (2005)
www.springerlink.com

from almost anywhere 24 hours per day, a Virtual "Me" could be created. Hence, the framework has been coined the "Me" framework.

Within this Me framework, content is structured using two models. The "Sniplet" model that gives structure to the knowledge and a "Multimedia Object" (MMO) Model that allows related multimedia to be packaged with meta-data utilizing a proposed Media Vocabulary Markup Language (MVML).

The chapter will also provide a brief look at some of the prototypes that have been implemented and trialed to test proof of concept and associated observations.

8.2 Managing Personal Knowledge

"Learning is expanding our personal knowledge"

Throughout our lives we accumulate knowledge. Methods used to expand this knowledge are incredibly diverse and can include instruction, research, discussion, experience or feedback. A common way in which individual learning is managed is through the creation of a personal portfolio of facts, information, knowledge and insights. Traditionally this has been achieved through summarising content, discussions, observations, and/or organising them in a paper based or electronic system. With the huge explosion of content available and the ease of content collection, particularly on the World Wide Web, techniques need to be developed to help manage personal learning content.

8.3 Personalised Electronic Content Management

Consider the case of students expanding their knowledge through instruction or research. Whether they are in face-to-face, virtual or blended environments, there is a need to have a mechanism that manages and catalogues their increasing knowledge. Resources that have been reviewed need to be tracked and in many cases annotated. Traditionally this would be done by annotating the original or a copy using a highlighter, with pencil in the margins, or summarising onto a separate sheet of paper or in an electronic document. For example if an article is deemed useful to be saved, it would need to be either copied, or a reference to a persistent copy would need to be recorded. This could be done by recording the URI (electronic), purchasing the book or writing the library reference. Then the article would be annotated to give it context with relation to the research or study.

At present this is a time consuming task. If the book/article is catalogued in an electronic database, such as the ACM digital Library (http://www.acm.org/portal/), some of the data can be extracted readily; in the ACM portal this is via BibTex Metadata. There is a lot of relevant and related material on the Web, in discussion forums or sent via email, but unfortunately the process of classifying and organizing such resources is often very time consuming. It would be a big advantage if the capture of this information could be simplified.

Content management systems exist for large content repositories, for example HyperWave, and tools are available for researchers to assist in the construction of literature reviews, for example EndNotes [4], but a real gap exists in the ability of individuals to easily capture the knowledge from the source and manage portfolios of their knowledge.

Therefore the question posed in this chapter is "can a personal content management framework be created, where an individual has the ability to create and share personalised electronic portfolios of their knowledge?" This includes:

- adding facts and information from research,
- managing personal knowledge and insights,
- providing the ability for others to review the knowledge (with appropriate permissions) and allow for their annotations,
- providing the ability for others to easily add your knowledge to their personal knowledge portfolios, and
- manage the multitude of ways the information can be represented electronically, and store or reference them so that they are saved in a persistent way.

8.4 Information and Knowledge

In the pursuit of increasing personal knowledge, information is synthesised from many sources. Unfortunately, for most, there is a problem with recall so mechanisms need to be developed that help us store the knowledge in a way that provides easy retrieval. Students use techniques such as keeping journals, making summaries, drawing concept maps and so forth.

This can be further generalized into the provision of information. A lecturer's knowledge is captured and used as the basis of the content to be delivered to students, which is then shared in lessons.

There is some debate as to what is knowledge and how this relates to information. It is useful to place knowledge and information in a pyramid as shown in Figure 8.1.

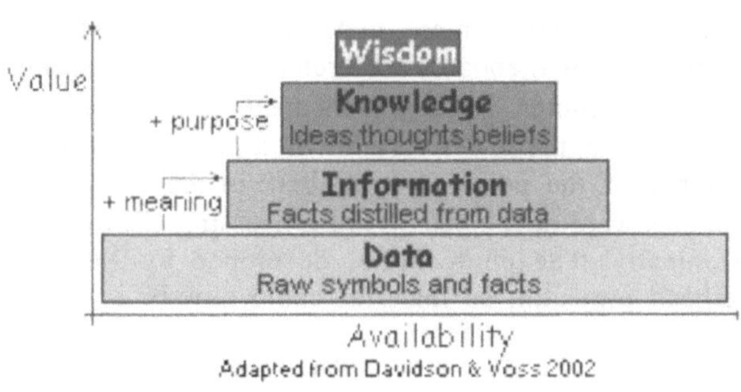

Fig. 8.1. Data-Wisdom pyramid [2].

The realization that knowledge capture is a two-way process creates many opportunities for collecting the knowledge. The goal then, is to develop a model that allows knowledge transfer in many directions, lecturer to student, student to student and ultimately student to lecturer. Indeed generalizing the model should enable multi-way knowledge sharing which is after all what occurs naturally.

The internet provides us with an electronic way to share knowledge. Current techniques include sending emails, participating in discussion lists, contributing to wikis to name a few.

The question is: Can a model be developed that more closely simulates what actually occurs in a social-human sense, where knowledge is "what we know", and is "what we share"? To be able to create such a model, requires a system that is accessible to all, can be structured in a way that the data integrity can be managed and uses technologies that are readily available.

8.5 The Personal Content Management Framework

With the rapid adoption of Web technologies, its ubiquitous nature provides the technologies that enable us to realize the framework. Progressively people are becoming “connected” and are using the web as a personal storage space. Google (http://www.google.com), for example have recognized this trend and in 2004 began trials on an email solution, GMail with its huge storage capability for each user. The Microsoft Network (http://www.msn.com) have created communities where individuals can organize social sites containing a variety of personal information and discussion networks. Individuals have also been managing and maintaining public web sites (such as, http://www.yahoo.com) to share personal information.

8.5.1 A background to Personal Content Management Systems

In order to answer the question “can we create a personal content management framework, where an individual has the ability to create and share an electronic summary of their knowledge?”, it is useful to consider how personal content has been managed in the past.

Prior to the introduction of electronic content, material was delivered through chalkboard to learner (felt-pens and whiteboards are a relatively new technology!). Books were expensive, so an instructor would present material in “chalkboard” chunks and this method was the main delivery method throughout the 20th Century. With the emergence of the personal computers in the early 1980s , several electronic technologies were developed. Word processing allowed textual data to be stored electronically. The photocopier, developed by Xerox also emerged, as did the overhead projector, so material could now be reproduced, displayed and stored more easily. This led to presentation tools, such as Microsoft PowerPoint and coupled with projection pads mounted on top of an overhead projector allowed for electronic presenting. The early 1990s saw the Static Web emerge and the spread of readily accessible networks with reasonable performance. The growth of on-line content changed the way research is done as "knowledge" can be obtained quickly and easily from a computer connected to the Internet pretty much anywhere in the World. A significant time investment was required for authors that maintained static pages. For example, one of the authors of this paper built a static content web to assist teaching delivery that had in excess of 4,000 html files and over 8,500 electronic media and multimedia files [11]. In order to manage and share the content, a two-layered model was developed where presentation included an A4 printable page and a form suitable for projection onto an overhead screen in a lecture. Figure 8.2 shows examples from the system.

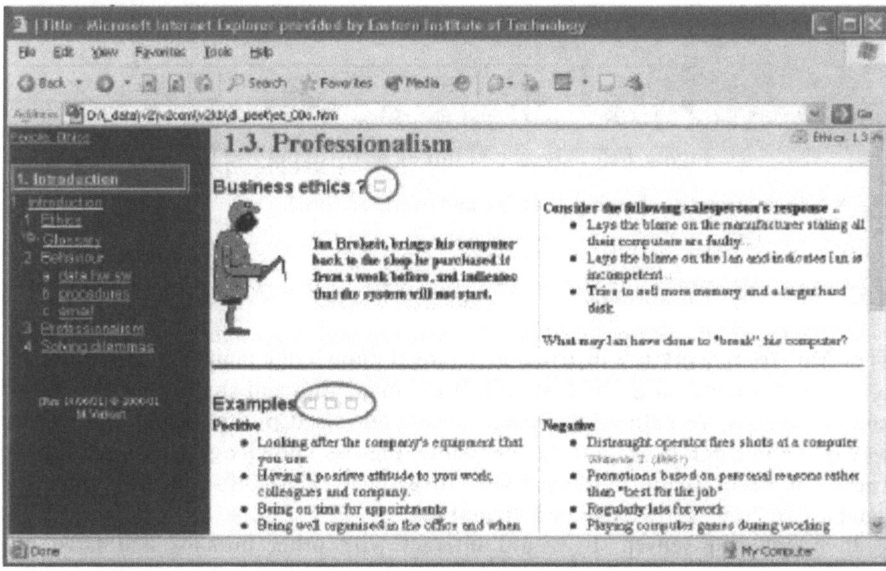

Figure 8.2a: Static web page showing hypergraphic to overheads

The development of web based database technologies in the early 2000s allowed interactivity to be developed, and the ability to manage more easily additions and modifications. A sniplet could now be represented as a database entity with related objects such as media and references. A prototype was implemented and tested over a number of years and from this research the Me framework emerged.

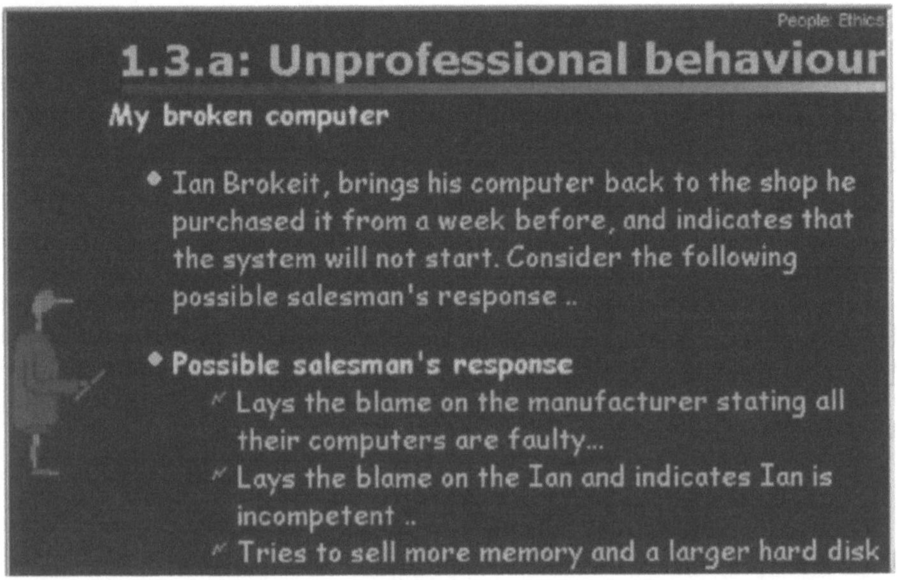

Figure 8.2b. Overhead linked to by hypergraphic

Fig. 8.2. Static web site showing printable and overhead forms.

8.5.2 The Me Framework

The "Me" framework is a distributed personal knowledge management model and is centered around an individual collecting, managing and sharing their personal knowledge. As we gather knowledge it is synthesised in a personal and unique way. If knowledge is to be truly accessible, then as knowledge is "something we know", it needs to be individualised. In essence, a Me framework contains a database of "my" knowledge or a **virtual me**.

In a physical sense, "I" would interact with other humans and have my knowledge expanded, modified or corrected. In order to have a system that captures knowledge the virtual me needs to have this ability also, and this can be achieved through the use of annotations. Initially it was envisaged that "external" individuals would be able to add core knowledge in the model, but with reflection this was discarded as a personal knowledge system needs to be moderated by the individual whose knowledge is being captured.

How then can the Me framework be implemented by a computer system? There are two necessary design separations, a logical model, which is how we see the framework and a physical model which is how this could be implemented in a computer system (how it is arranged on the disks and in the files).

8.5.2.1 Logical Model

The logical model view of the Me framework is a diagram drawn from the perspective of how a user would see the system, rather than the physical view, and is illustrated in Figure 8.3.

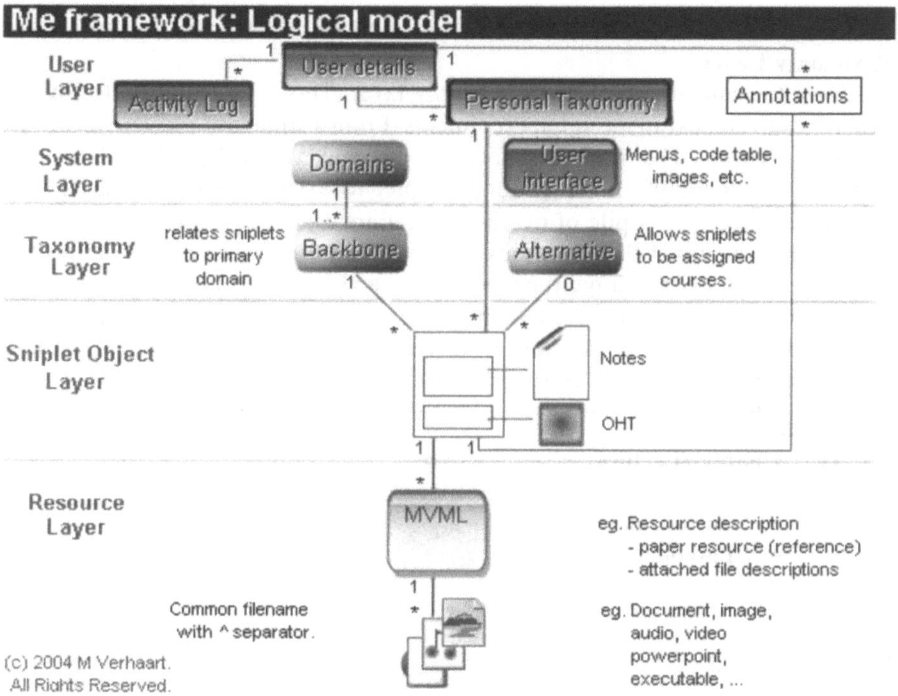

Fig. 8.3. Me Framework: Logical model

The logical model contains the following layers.

Sniplet Layer

The core of the model is the actual knowledge itself. In order to refer to a fragment of knowledge the term "**Sniplet**" has been coined, and is defined as "**a piece of knowledge or information that could be represented by one overhead transparency**" [12]. In essence this means that the concept can be explained succinctly using electronic media such as text and images. Text suitable for both an overhead and supporting background notes are contained in the sniplet. This is consistent with the current literature on learning objects. The sniplet layer and its relationship to learning objects will be discussed later in the chapter.

Resource Layer

Attached to each sniplet is a collection of resources, or *digital assets*. These include links to web based resources, multimedia elements or references to physical resources such as books or journals. In order to describe and manage the individual resources a Multimedia Object (MMO) Model has been developed, with an associated XML based vocabulary, coined the Multimedia Vocabulary Markup Language (MVML), that provides the meta-data to the resource. The structure and importance of this will be described later in the chapter.

Taxonomy Layer

Above the sniplet layer is the Taxonomy layer. This is essential as it provides structure and context to the sniplet fragments. From a structural view there are two taxonomy views. The first is what Guarino and Welty [5] referred to as a "Backbone taxonomy". In order for a sniplet to exist it needs to placed in a context, for example, a pile of overheads that are randomly shuffled are of little use. Further, a sniplet that describes how a digital photo in a jpg file is resized would probably exist in a multimedia domain and in its taxonomy. But, a sniplet may need to exist in an alternative context, or taxonomy. For example, the resizing of a jpg file may also be relevant in a web design domain or even a database domain where images are to be stored efficiently.

Domain
- **Chapter/Lecture**
  - **Section**

Internet: Structure and Use
- Introduction
  - Objectives
  - Overview
  - ....
- History & Future
  - ....
- How the Internet Works
  - ....

Web Markup Languages
- ....

Multimedia
....
Macromedia Director
....
Macromedia Flash
....
Databases
....
MS-Access
....

Fig. 8.4. Example of partial backbone taxonomy of domain of Internet

So the Me model allows for each sniplet to be assigned to a backbone taxonomy and to exist in alternative taxonomies. Within the taxonomy a sniplet may have an association with another, for example, it may be part of a sequence of sniplets, or may be a sub-category of a sniplet. A tree structure is probably the simplest structure that will allow sequencing and sub-sectionalisation.

An example of partial backbone taxonomy that could be used in a lecturing scenario in the Internet domain is shown in Figure 8.4.

Figures 8.5 and 8.6 show a web-based prototype that was developed to illustrate how this could be implemented.

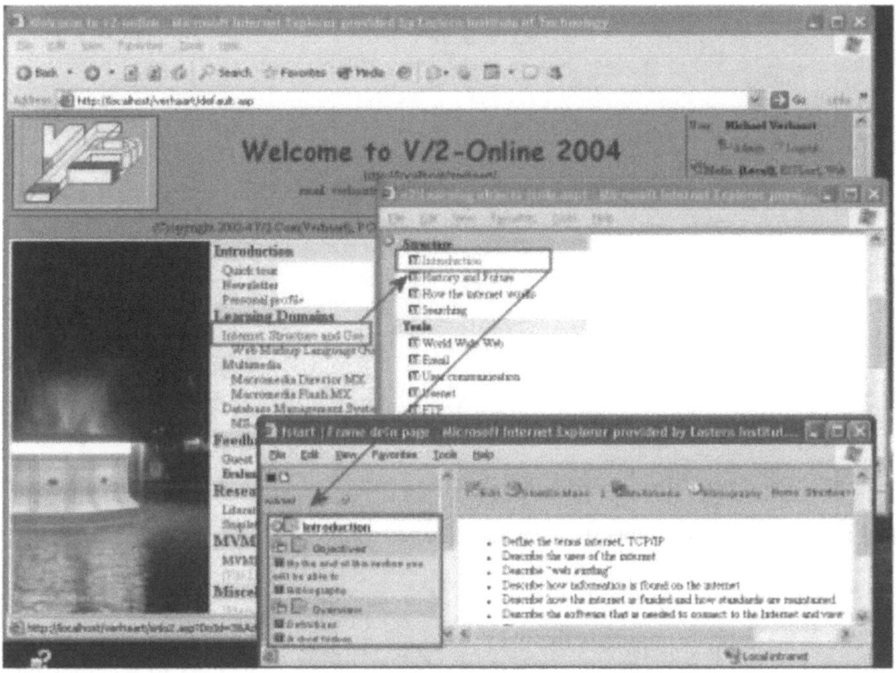

Fig. 8.5. The taxonomy of the sniplet model

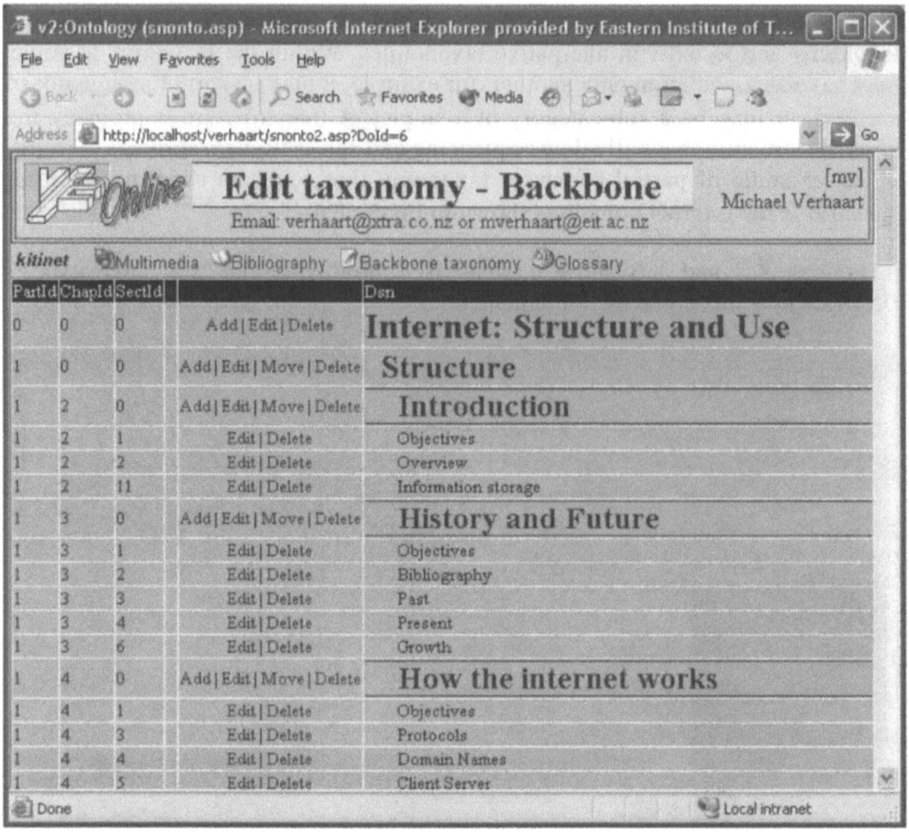

Fig. 8.6. Taxonomy as a table showing maintenance options

System Layer

The system layer manages the user interface. In Figure 8.3 content domains are specified at the system level. At present there is some debate as to whether this should remain at the system level or be moved into the taxonomy layer. From a physical organisation view, separating domains allows for more efficient management of resources as resources in a particular domain are easily grouped together. A major drawback on separating the domains at system level, which is how the current prototype has been constructed, is the inability to easily share content between domains and secondly to allow for global search-ability in through all the domains.

The second part of the system layer are the settings and codes required by the whole system. This includes the ability to dynamically manage a site menu. The site menu needs to link to domain and alternative taxonomies, to allow for menu headings (improves menu navigation), and the ability to link directly to URLs. Linking directly to a URL has been found necessary to enable the inclusion of pages such as a "Personal Profile" page, and to other software such as on-line evaluations. The system layer also manages global codes. These include the templates for the referencing styles. For example an APA style reference for a Web resource with a known author is:

&c (&d|n.d.|) &t. Retrieved &r, from &w

where

&c = creator,

&d = date the | gives a choice if the field is blank (displays n.d. no date),

&t = title, &r = retrieved date and &w = url.

One of the features that improves the usability of the Virtual Me is the ability to assist users where English is a second language. A translation table can be made available for specific words and these can be identified by the system when the sniplet is displayed. Valid language codes can be stored in the code table. Default values can also be maintained in the code table, such as the default Cascading Style Sheets to be used when a new user is created.

Alternative site access for multimedia is also managed at the system level. The prototype identified that there is an issue associated with the location of multimedia. The logical place to store the "Virtual Me" would be on the Web, and would include both the html pages and the multimedia files. Unfortunately it was found that some corporate firewalls restrict the availability of some media types, such as audio and video, and in more extreme cases images. So the ability to switch where the multimedia can be sourced is important. In this corporate scenario, the multimedia can be stored on a local file server while the html files can be accessed from the web site. In a teaching environment many students may access web sites using dial-up access, or access that is paid for by the megabyte. In this case students could be supplied with a CD-ROM containing the multimedia, and reference the web for the html files. One drawback of this approach is that as the html files are dynamic, the multimedia on the CD-ROM could become obsolete or not be synchronized with those on the web server.

The system layer can also allow for personal customization of the "Virtual Me". For example, a collection of startup images could be managed at the system level, or a startup message for the home page of the web site, that is displayed prior to the user logging on.

User Layer

The user layer manages user settings. Figure 8.7 illustrates the type of data that could be captured to manage a user profile.

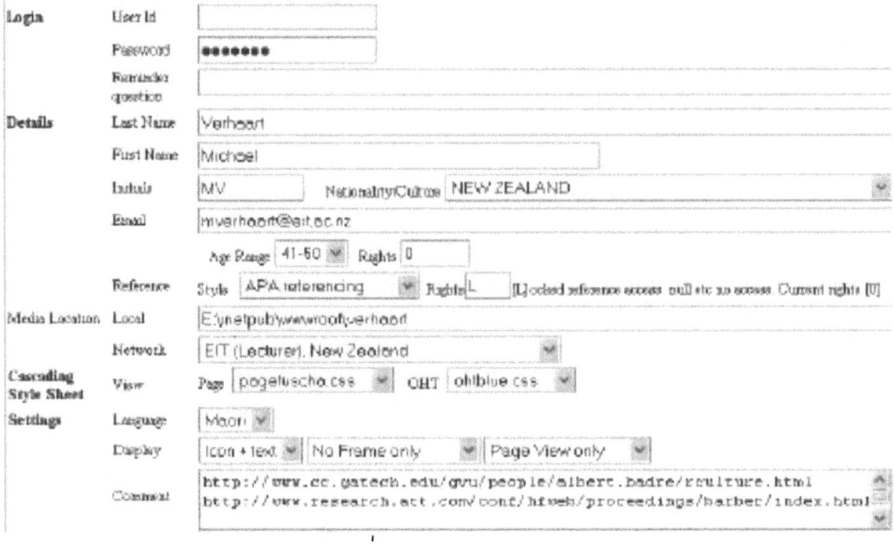

Fig. 8.7. Model of user settings

Aside from the common fields that include the user's identifier, password, name, email address and so forth, the comment field is an important one for the Virtual Me. In the Me framework it can be used as an alternative communication medium that allows one-to-one comments to be made between the user and the owner of the Virtual Me. This is consistent with a physical me where a person communicates directly with the knowledge owner. Also in a physical world, a person would introduce themselves then gather the knowledge they were seeking. Similarly, the Virtual Me allows new users to create their own profiles, facilitating a simpler management of the personal online data capture. Experience from an actual prototype demonstrated the need for the reminder question, as users frequently forgot their passwords. It was found that users choose one of two solutions to resolve this problem; the first is the user creates another new profile or in the second case emails the owner of the Virtual Me to change their password.

The user layer also manages a user activity log where movement within the system and frequency of access is monitored.

Annotating content has become increasingly common on the internet. Web pages containing news articles solicit user feedback (e.g. http://www.newsfactor.com , http://www.askmecorp.com/). Indeed W3C in the Annotea project (http://annotest.w3.org) are developing web server solutions that will manage annotations. In the context of this research, this is probably the most important feature, enabling the capture of personal knowledge that allows for the creation of a virtual learning portfolio. Many content management systems rely on bulletin boards or email systems to manage annotations. The biggest drawback of these systems is that the annotation is separated from the content. This separation often results in the actual annotation not being applied to the content. For example, an email is sent indicating an amendment is required to the content.

From an authoring view, this means that the content has to be located, and in the case of thousands of pages this more often that not doesn't take place. Allowing others access to our personal portfolio is a small step to sharing knowledge, but allowing them to contribute their knowledge to the portfolio provides the real power to increase the stored knowledge. In the Me model the annotations are attached directly to the sniplet thus retaining the correct context for the comment. The ability of the owner to accept or reject the annotation is important as is the ability of the annotations creator to edit or delete it.

8.5.2.2 Applications of Me Framework

In order to understand the rational behind the Me framework it is useful to consider possible applications. This section briefly describes some use cases that show the benefits of adopting the framework.

As a teaching delivery tool

A lecturer records their knowledge in their virtual learning portfolio and it is organised into a backbone taxonomy. If necessary, an alternative taxonomy can be created that is specific to the lesson to be taught. Students are granted access to this content. Since the Me framework uses a "Sniplet" which can be displayed as an overhead projection, the content can be delivered as a lecture. Further, as the sniplets are available as notes these can be printed or reviewed by students before or after class. Students can increase the value of the content by adding their knowledge via annotations and these can be incorporated into the content by the owner of the portfolio.

If the student also had a personal portfolio, they could capture the lecture material and reorganize it in a form suitable for their recall.

Managing research

Research requires considerable reading and data collection. Articles, books, journals, web resources are all combined with personal knowledge to produce a literature analysis. Using the Me framework would allow for the management of these resources and the ability to add personal notes. Further the use of annotations allow for ad-hoc observations to be added to the reference plus the ability of others to review and annotate the references. The use of the notes feature and an ability to produce a sequential printout based on the research reviews taxonomy, would allow this research to be published in a draft format. A more detailed discussion of a literature review using the Me framework will be covered later in this chapter.

Sharing personal information

A person's knowledge is captured in the virtual portfolio. This includes personal images of their family, and possibly observations that have been made.

Access to family and friends is given to enable them to view this information. An annotation feature would allow them to provide feedback regarding the content.

Business contacts can be maintained in the virtual portfolio. With standardization and as shown later, the use of the vCard standard to describe the creator of a resource, the ability to manage personal contacts can be simplified. For example, a business card manager could easily be constructed.

8.6 Core Models in the Me Framework

With the Me framework described, the two core models will now be covered. These being the sniplet model and the Multimedia Object Model (MMO).

8.6.1 The Sniplet Model

The Sniplet model evolved from a need to represent and organize knowledge in manageable chunks. In the context of a teaching situation, a piece of content is normally presented to the class as an overhead projection (or as a whiteboard chunk), and is supported by lecture notes. These fragments can then be organized into a lesson, and a group of lessons form a content domain.

Based on this requirement, the following sniplet specifications evolved.

- Metadata describing the sniplet; such as; a mandatory title, a selection of dates, for example; creation date, flag date (if a flag should be raised for this sniplet on a particular date) and deletion date.
- Textual content for both a description (for the printed material) and a summary (for the overhead).
- Digital Assets
  - Multimedia attachments. This would include unstructured media elements (text, image, animation, sound or video) or structured elements (for example, PowerPoint slide set, Portable Document Format file (PDF), Flash movie, customized software).
  - Bibliographic attachments. As knowledge is usually based on a citable reference it is necessary to provide the ability to attach a citable resource.
- Annotations. Comments can be attached to each sniplet. The annotation meta-data must include a sniplet and a user identifier, as well as the creation date. An action factor would be included to indicate if there is any follow up required by the owner of the portfolio, and would be in the range "urgent" to "for your information". An initial design included a rating of the associated sniplet, but in a test prototype this was seldom used. Some analysis fields can also be included such as the type of annotation (for example; Social, Information, Knowledge), whether the annotation can be archived and if the annotation should be private between the user entering the annotation and the owner of the portfolio.

To prove that the model was feasible, a prototype was built, which provided a learning environment to test suitable web technologies. This is described in detail by Verhaart & Kinshuk [13], and a sample sniplet from the prototype is shown in Figure 8.8.

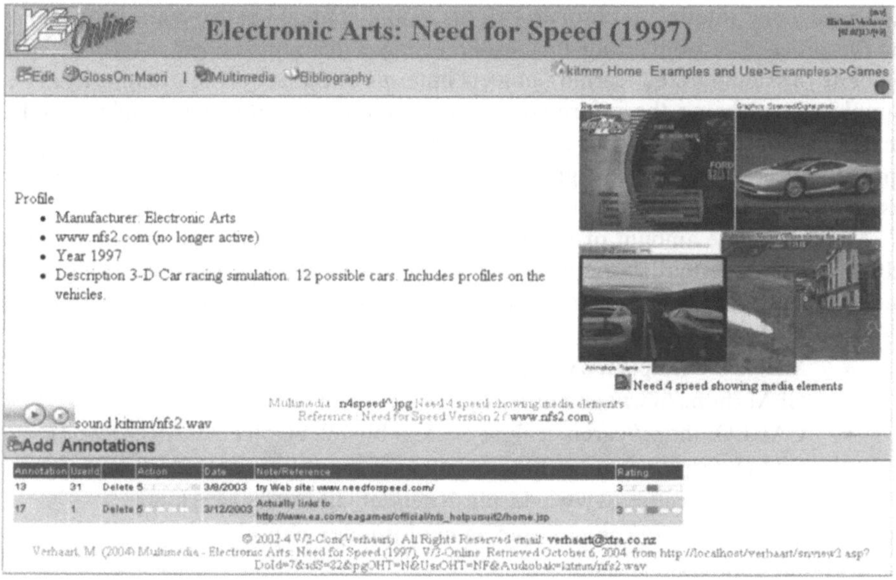

Fig. 8.8. Sample of a sniplet prototype showing attachments including an image, a sound and two annotations.

Figure 8.8 illustrates several of the sniplet design features . The title "Electronic Arts: Need for Speed (1997)" is an example of the sniplet metadata. Textual information is minimal and just gives a brief profile of the game. Digital assets include the collage of screen captures, a sample sound file and a link to the associated web site. The two annotations were added by users (shown at the bottom of the screen) as one of the users found the original web site URL had changed.

As the main context that the framework is being developed in is an educational setting, it is relevant to explore what similar technologies are being used in this environment.

8.6.1.1 Relationship of Sniplet to Learning Object Research

A need to develop learning resources that could be reused and shared on the internet has led to the development of what are now commonly referred to as learning objects (LOs). An early specification was published by IEEE in 1999 [7] and provided a very generalized definition, and that included persons, organisations and events. Most current definitions now include a pedagogical element.

For example, DLNET [1] defines a learning object a structured, standalone resource that encapsulates high quality information in a manner that facilitates learning and pedagogy. It has a stated objective, a designated audience, ownership and associated intellectual property rights, so its content shall remain unchanged in the process of converting the resource into a learning object. McGreal and Roberts [8] summarized learning objects in terms of their level of granularity. The simplest level being the information object or *component* (for example, simple text, a photograph,) and these can be attached to a *lesson*. Lessons can be grouped into *modules* and modules aggregated to *programs*. So learning objects can exist as *components*, *lessons*, *modules* and as a *programme*.

A considerable amount of money is invested in content development, and the ability to develop collaboratively and share this content is possible using web technologies. There are many initiatives in this area, from learning object repositories such as

- MERLOT (http://www.merlot.org),
- GEM (http://www.geminfo.org),
- CAREO: (http://careo.ucalgary.ca/cgi-bin/WebObjects/CAREO.woa),
- LALO (http://www.learning-objects.net),
- Edusplash (http://www.edusplash.net/default.asp?page=Home),
- VCILT (http://vcampus.uom.ac.mu/lor/index.php),

to digital libraries such as

- the Greenstone project (http://greenstone.org/),
- DSPACE (http://www.dspace.org) and
- the ACM digital library (http://acm.org).

The development of learning objects has a direct relevance to building a personal portfolio as many of the technologies, structures and issues are similar.

In the analysis of the current state of Learning Object repositories Verhaart [14] concluded that LORs appear to have actually developed into portals, pointing to web content and indeed it is often better to use a standard search engine such as Google to find suitable teaching content.

So given that LO repositories are evolving into portals , this research proposes that the user centric model (Me framework) managing an individuals knowledge portfolio is the next evolutionary phase to distributed knowledge sharing.

There is ample anecdotal evidence that supports the development of personal knowledge management systems, for example, in the face-to-face teaching area, instructors customize content to suit personal delivery styles. There is also much current research in the area of adaptive learning environments, where content is presented in a "customized" way based on learner preferences. A *Blog* which is an online diary or journal, typically documenting the day-to-day life of an individual, and derived from "weblog"[10] has become a way that individuals can record unstructured comments and allow reader annotations.

8.6.1.2 Building a Personal Portfolio Using the Sniplet Model

As mentioned earlier an application that requires managing and organizing a significant amount of personal knowledge, involves researching a topic, annotating appropriate research, organizing the literature review in a way that allows preparation and construction for publication. The following discussion describes how the sniplet model can be used to manage that process.

Building the taxonomy

The first stage requires the development of a taxonomy to provide a framework. For example, Figure 8.9 illustrates a simplified extract of the taxonomy on capturing meta-data.

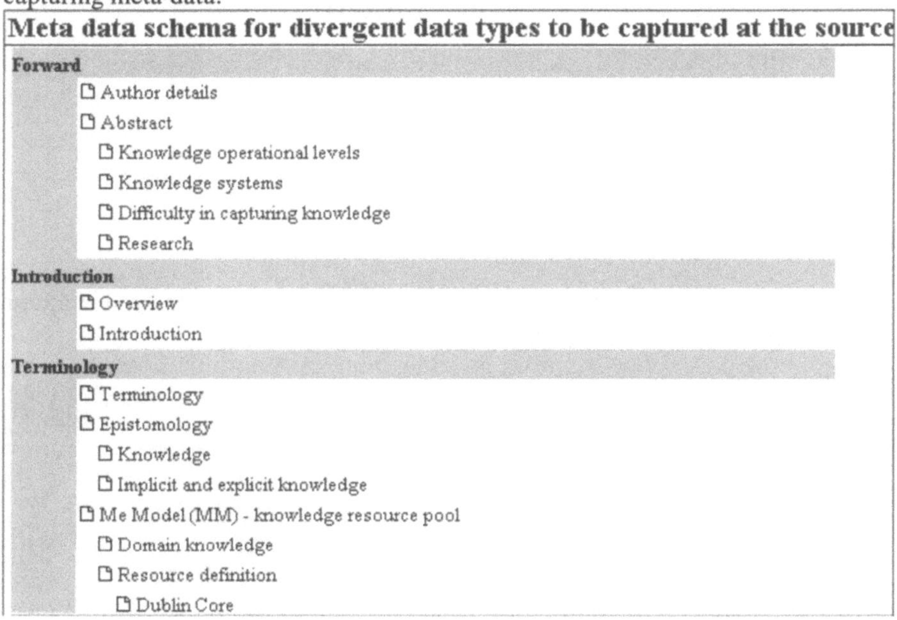

Fig. 8.9. Extract of a taxonomy for a literature review

Adding the sniplets

Once the taxonomy is created the sniplets can be added. Figure 8.10 illustrates a prototype model of a sniplet entry. Essential entries include the division, which is the link to the taxonomy, the title and the text description. The summary is used for the overhead projection, and if missing the full description would be used.

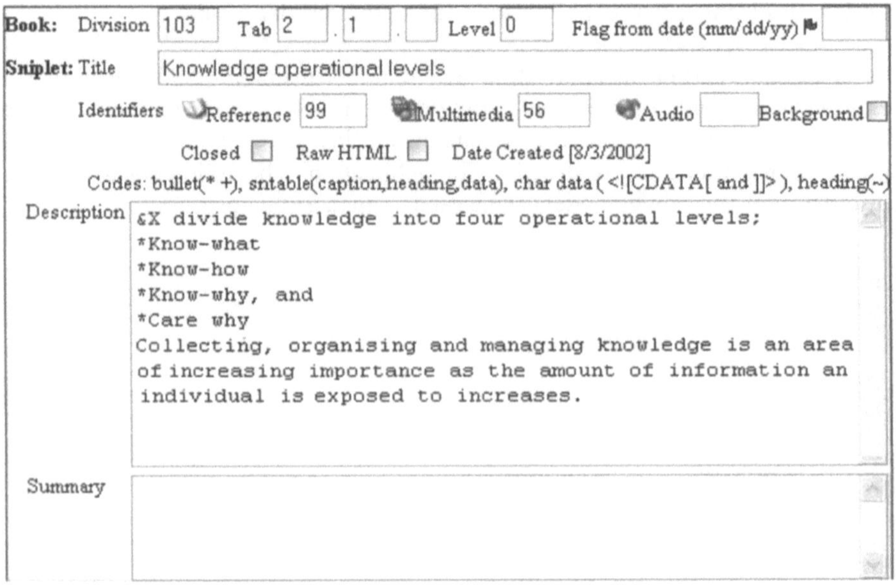

Fig. 8.10. Model of sniplet data

Once saved, the sniplet data shown in Figure 8.10 would be displayed in a form similar to that illustrated in Figure 8.11.

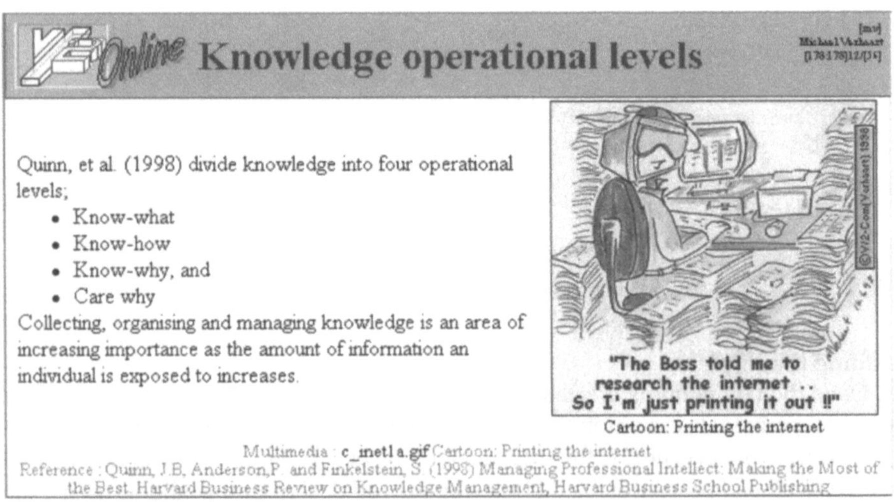

Fig. 8.11. Sample of a sniplet

Bibliographic management

Finally, and most importantly, the portfolio needs a method to manage the actual bibliographic material. Products such as the MS-Word add-in end-notes [4] provide some of the facilities especially in the area of both inline and bibliographic referencing. The ability to add bibliographic references is therefore essential, and in Figure 8.10 a reference identifier (value 99 towards the top of the screen) is shown. A possible structure for entering bibliographic data is illustrated in Figure 8.12.

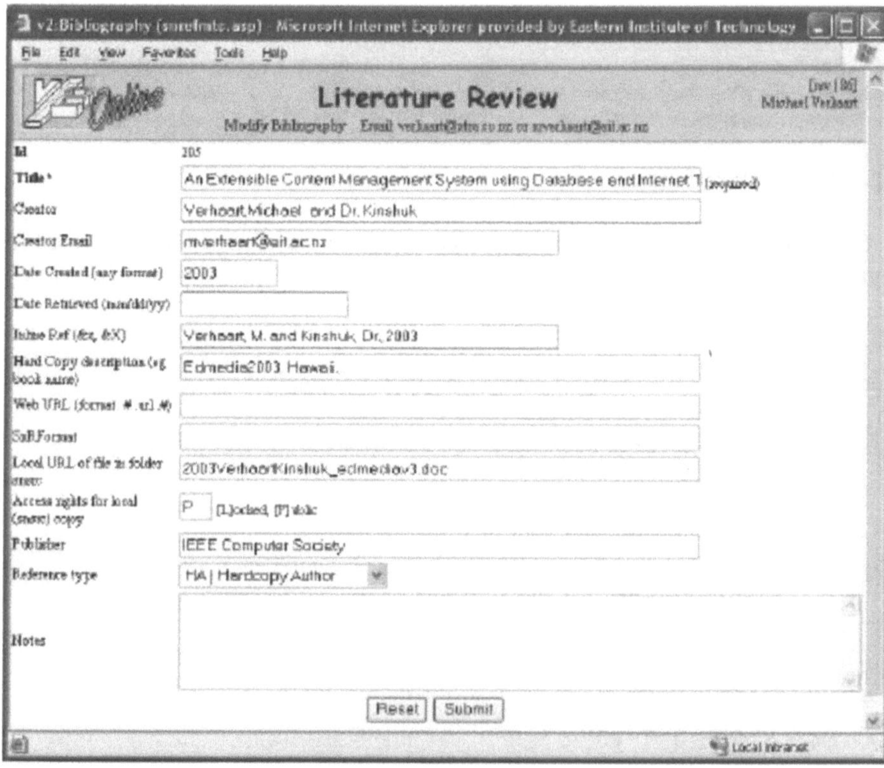

Fig. 8.12. : Metadata for a bibliographic reference

A key requirement is that the attributes (fields) making up the bibliographic meta-data must be re-constructible into a standard referencing format such as that from the American Psychological Association (APA). From Figure 8.12 an APA formatted reference would take the form

"Verhaart,Michael and Dr. Kinshuk (2003) [13] An Extensible Content Management System using Database and Internet Technology. Edmedia2003, Hawaii., IEEE Computer Society { P | 2003VerhaartKinshuk_edmediav3.doc }"

The ability to save and link to a local copy of the actual source document is a useful function, especially where it exists in an electronic format. For a web page this can be critical, as often these are moved from the original location. Google manages this by retaining an copy of all pages that are indexed in cache (Figure 8.13).

Fig. 8.13. Google cache page

Once the bibliographic data is captured it can be transformed as illustrated in Figure 8.14.

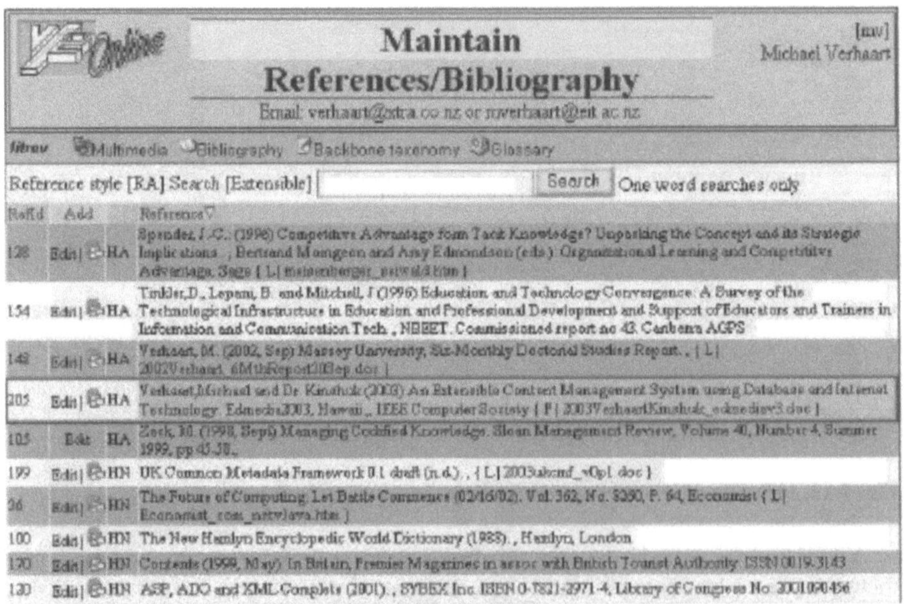

Fig. 8.14. Bibliographic table showing features

Figure 8.14 demonstrates the following features;

- The transformation of the data into an APA style reference. The use of a formatting string, discussed earlier, allows for style modifications.
- A link to a local file is shown in the {}, with a L (local) or P (Public) prefix showing access rights. {P | 2003VerhaartKinshuk_edmediav3.doc}
- The use of database technologies allows additional functionality including;

- o A search facility
- o Linked relationships to sniplets that use this reference. The icon to the right of "Edit" indicates that this entry is used by one of the sniplets (otherwise it is left blank, Ref 105) and clicking on the icon will display sniplets' associated.

8.6.1.3 The Structure of the Sniplet Model

In summary, the sniplet model has the following features.

- An association in a backbone taxonomy to give the sniplet context
- A possible association with alternative taxonomies. For example, a sniplet on multimedia could be relevant to other domains such as; Internet, Database, electronic design and photography.
- A mandatory title, and creation auditing data (such as date and author)
- A description and optionally a summary that is used to display the sniplet as an overhead. If the summary is missing the description is substituted.
- The ability to link annotations is key to the capturing of knowledge, both from the owner and of the users.

And optionally may contain

- Flags to manage deletions and permissions. For exa mple, the sniplet can be available only to the owner of the Me system.
- Dates to hide the sniplet until the date specified is reached.

The final related components are the digital assets linked to the sniplet, that is, the multimedia and bibliographic objects.

8.6.2 The Multimedia Object (MMO) Model

Many systems have been developed to cope with the classification of paper based resources, and indeed this is the principal job of a library. Historically, a card system and now electronic files are used to contain the meta-data that provided the link to a physical book or publication.

Much work has been done in this area, and from this the Dublin Core [3] meta-data scheme was developed. Basically the Dublin Core is a set of 15 fields such as title and author, and all are optional. Many electronic systems store the meta-data as a record in a database while the object it describes is either a physical resource or accessed via a link. Who creates the meta-data is the subject of much debate. McGreal & Roberts [8] define two camps for capturing meta-data in the context of learning objects.

"The “internal reference” camp believes that the creators of learning objects should input their own metadata. The “external reference” camp believes that only librarians or information specialists should input the data. The “external” camp argues that only professionals can ensure the integrity of the data that is input. The “internal” camp argues that the number of learning objects is growing so rapidly that there is simply too much work required, making it impossible to leave it to small numbers of professionals. They also argue that engaging professionals would be very expensive."

A good example of externally created meta data is the ACM digital library (Figure 8.15).

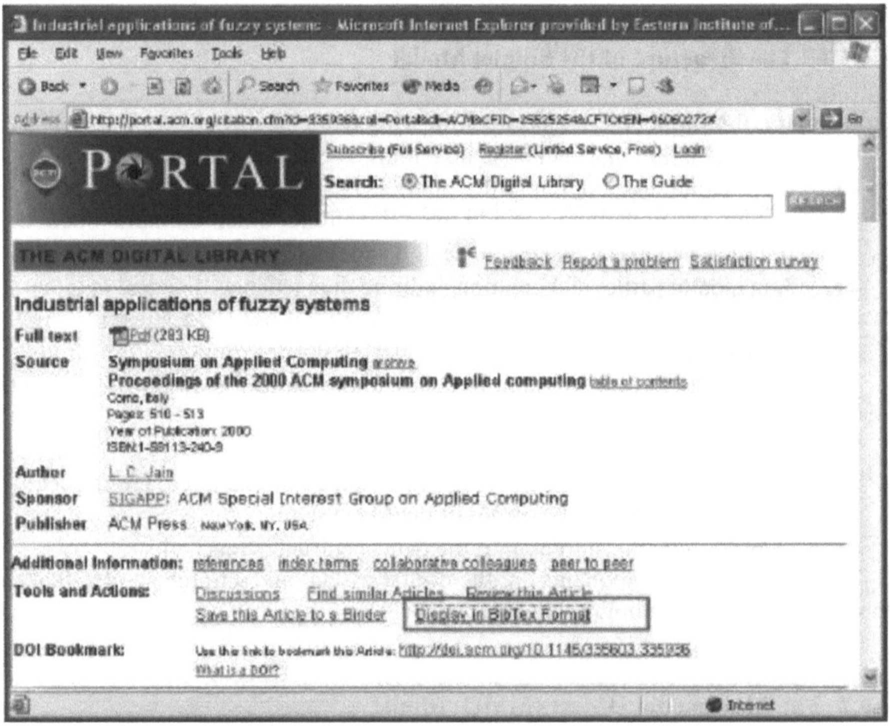

Fig. 8.15. ACM Portal showing a external meta-data

In the ACM digital library, this meta-data can be shared via a BibTex format. A detailed description can be found in Erik Heitfield's 1996 paper "Using BibTex and epsfig" [6].

For the paper described in Figure 8.15, The BibTex link (bottom right) produces the copyable meta-data illustrated in Figure 8.16.

```
@inproceedings{335936,
author = {L. C. Jain},
title = {Industrial applications of fuzzy systems},
booktitle = {Proceedings of the 2000 ACM symposium on Applied computing},
year = {2000},
isbn = {1-58113-240-9},
pages = {510--513},
location = {Como, Italy},
doi = {http://doi.acm.org/10.1145/335603.335936},
publisher = {ACM Press},
}
```

Fig. 8.16. BibTex Metadata

This is great for entries that are in a digital library, but unfortunately there are many resources that are not. But resources are not only in a document format, other media types may also be used as a source of knowledge. These include other media types, such as images, animation, sound and video.

So the question that needed to be addressed is "can we describe resources in a common way so as to allow for clear referencing, content description and ownership?". Verhaart, Jamieson & Kinshuk [15] described a meta-data language (MVML) that was designed to describe resources, and was based on several commonly used meta-data schemas.

8.6.2.1 MVML Files

A core part of the portfolio is the creation, management and organisation of **digital assets.** A digital asset is a computer file containing "unstructured" data, such as an image, a video or audio clip or "structured" data such as a document or spreadsheet that has been tagged with descriptive information [16], and were first classified in the late 1990s [9]. Typically the descriptive information is in the form of meta-data that is in some way attached to the computer file.

In the Me framework this concept has been extended from a single file with descriptive information to a related collection of files (a manifest) and a metadata file [15],. Many web based systems use a thumbnail of an image with a hypergraphic link to the actual image. This improves download speed and minimizes bandwidth requirements. So in essence there are multiple occurrences of a file that can be displayed, and the form displayed will depend on the context. Further, an image may require other forms, for example, if a user is blind a textual description for use by a screen reader could be attached, or alternatively a verbal description. So a digital asset may exist as an image file, and iconic image file, a sound file describing the image file and so forth. These are then described in an eXtensible Markup Language (XML) meta-data file using a Media vocabulary Markup Language (MVML).

The term Multimedia Object (MMO) was coined to describe the collection of files and the associated MVML file that describe a digital asset.

In order to be a viable way to manage digital content several issues need to be considered. Firstly, creation of an MMO needs to be straightforward and simple. It is commonly felt that the success of HTML as a markup language is due to its simplicity, and indeed keeping the MMO creation as straight-forward as possible is essential in its design.

Secondly, any meta-data needs to conform to current practice and standards. So, common standards have been adopted to describe the object and these have been incorporated into MVML. XML has been steadily gaining momentum as the future of the internet, and indeed is the cornerstone of the Semantic Web. Probably the most widely used meta-data standard is the Dublin core, and can co-exist comfortably with other meta-data sets. [8]. Unfortunately, there are no required fields which makes structured computer processing and interpretation difficult, and so, MVM L requires a title as a minimum entry. Other standards that have been incorporated into MVML are

- the Resource Definition Framework (RDF), which is essentially a way to uniquely identify a resource whether electronic or not.
- vCard, which is a meta-data format that enables a person to be described. This is used extensively in commercial email systems and can be thought of as an electronic business card.

Unfortunately, trying to include these meta-data standards into a common scheme increases complexity. So, a way to make entry of the meta-data by an end user as simple as possible is essential to give the model any chance of success.

To illustrate how an MVML file could be created by an end-user a workstation prototype was developed, and the basic meta-data entry is illustrated in Figure 8.17.

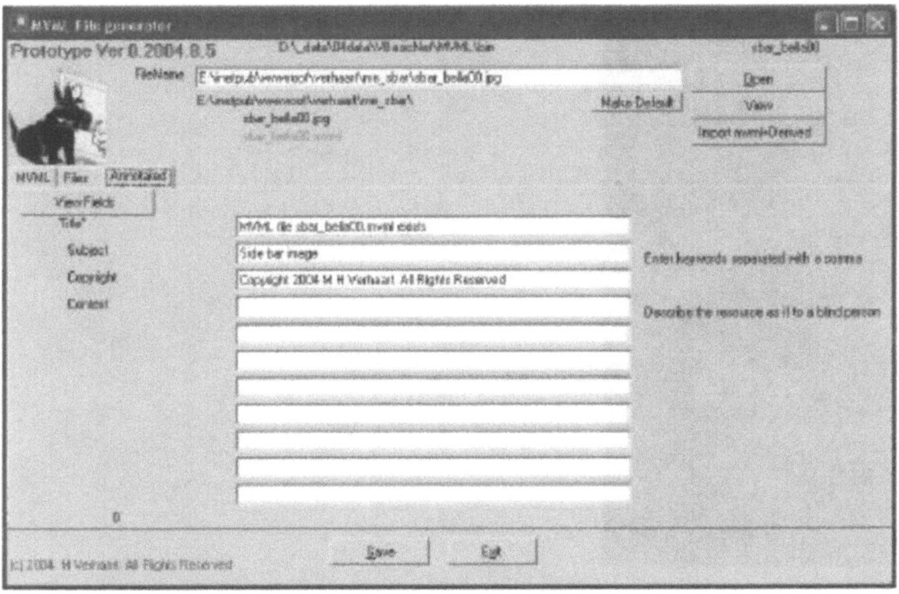

Fig. 8.17. Creating an MVML file

Although from the screen shot this appears to be simple, on creation a considerable amount of meta-data can be generated. This is achieved via a combination of automatic capturing of derived meta-data and a template. Derived meta-data can be extracted by the computer and for an image would includes things such as its size, type and creation date.

The template is the most important thing enabling quick creation of annotated meta-data. For example, as the Me framework is essentially the creation of a personal space, the creator of any items within this space will be "me". So the creator information can be entered into the template and this can be automatically included in the MVML file attached to the digital asset.

An example of an MVML file attached to an image is illustrated in Figure 8.18.

Figure 8.18a: Image of Scottish Terrier (Bella).

```
?xml version="1.0"?>
 <rdf:RDF
  xmlns:rdf  = "http://www.w3.org/1999/02/22-rdf-syntax-ns#"
  xmlns:dc   = "http://purl.org/dc/elements/1.1/"
  xmlns:vCard = "http://www.w3.org/2001/vcard-rdf/3.0#"
  xmlns:mvml = "http://is-research.massey.ac.nz/verhaart/1.0" >

 <rdf:Description rdf:about = "sbar_bella00.mvml" >
  <dc:title>Bella, a scottish terrier on a tree trunk</dc:title>
  <dc:creator rdf:parseType="Resource">
      <vCard:FN>Michael Verhaart</vCard:FN>
      <vCard:N rdf:parseType="Resource">
        <vCard:Family>Verhaart</vCard:Family>
        <vCard:Given>Michael</vCard:Given>
      </vCard:N>
      <vCard:EMAIL rdf:parseType="Resource">
        <rdf:value>mverhaart@eit.ac.nz</rdf:value>
        <rdf:type                rdf:resource="http://www.w3.org/2001/vcard-
rdf/3.0#internet"/>
      </vCard:EMAIL>
  </dc:creator>
  <dc:date>2002-11-01</dc:date>
  <dc:source rdf:parseType="Resource">
   <mvml:URL>
     <mvml:root>http://is-research.massey.ac.nz/verhaart/</mvml:root>
```

```
        <mvml:path>me_sbar/</mvml:path>
        <mvml:file>sbar_bella00.jpg</mvml:file>
      </mvml:URL>
    </dc:source>
    <dc:title xml:lang="fr">LScottish Terrier on le Tree Trunk</dc:title>
    <dc:subject>Scottish Terrier, Bella, Dog, Tree stump</dc:subject>
    <dc:rights>Copyright 2002. All Rights Reserved</dc:rights>
   </rdf:Description>
   <mvml:manifest>
    <rdf:Description rdf:about = "sbar_bella00.jpg" >
      <mvml:width>200</mvml:width><mvml:height>400</mvml:height>
      <mvml:type>IMG</mvml:type>
      <mvml:context>A photo of a black scottish terrier called Bella standing on
the trunk of a cut down redwood tree stump in Havelock North New
Zealand</mvml:context>
    </rdf:Description>
    <rdf:Description rdf:about = "sbar_bella00^icon.jpg" >
      <mvml:width>47</mvml:width><mvml:height>94</mvml:height>
      <mvml:type>IMG</mvml:type>
        <mvml:context>A small photo of a black scottish terrier called Bella
standing on the trunk of a cut down redwood tree stump in Havelock North New
Zealand</mvml:context>
    </rdf:Description>
   </mvml:manifest>
  </rdf:RDF>
```


Figure 8.18b: Example of an MVML file attached to an image collection.

Fig. 8.18. Example of an image and associated MVML file

The proposed model contains the following features.

Firstly, an MMO can be completely contained within an MVML file. This is the case where plain text is the only media object required.

Secondly, and this would be the more common case, an MMO would consist of an MVML file and one or more associated multimedia files. To allow for grouping and human recognition, the MMO has a standard file naming convention. The principal object and the MVML file have the same name, but different extensions, for example; myfile.mvml and myfile.gif. If there are additional files they are either given the same name with an alternative extension or given a ^ suffix. So in the example if there were a sound file it could be named myfile.wav, or if an icon file were available it could be myfile^icon.gif.

There is some discussion at present as to whether the MMO should be a zip file, but regardless it is intended that the file name conventions be adhered to.

As to existing multimedia meta-data standards such as Apple's mov format, or the Motion Picture experts group (mpeg) formats. The authors believe that these proprietary standards are not portable between the wide variety of existing systems and require specialized software to interpret the results. The proposed format works for all types of multimedia and will cope with any legacy and future ones.

8.7 Conclusions and Future Directions

This chapter has discussed a framework for creating a personal learning portfolio. Standardisation of content fragments into the "sniplet" format, and digital assets into multimedia objects (MMOs) using a self describing meta-data language (MVML) will allow for portability and ease of sharing. The learning portfolio is basically a collection of sniplets created by the portfolio's owner. The owner can either add new sniplets of knowledge or add annotations to existing sniplets. In the real world knowledge can be expanded by others we have contact with. In the virtual environment there is an opportunity to allow visitors to our personal learning portfolio to add their own knowledge, by giving them the ability to annotate in context. Annotations can be attached to either a resource or to a sniplet. If the annotation is to be added to a resource (digital asset), this should be added directly to the MVML meta-data file. Sniplet Object annotations, allow for "knowledge" to be added to the portfolio in context. Initially it was thought that sniplets would be created by individuals that were given access to the system. This has been shelved in preference to annotations and allows for integrity issues to be addressed.

So, the Me model is about creating a virtual knowledge portfolio for an individual. In many ways this is contrary to the current trend of creating large repositories of learning objects. In the Me repository, a variety of information is stored, from personal files such as a photo library to sharable files such as teaching content, and if it were to be adopted as a standard, individuals could share digital assets and knowledge easily. For example, if a person wanted a media element, copying the MMO (files plus MVML meta-data) would automatically provide all the contextual data with the media element. A bibliographic object (for example, web site or research article), would have attached bibliographic meta-data so that automatic referencing becomes possible. From a research view this would produce a significant time saving. A second advantage is that each digital resource would come complete with ownership and copyright details in a form that is easily accessed. There are tertiary institutions, that require accurate referencing of digital assets before they can be used in lectures. The University of Melbourne has such a policy, and was described by J. Pearce in a personal communication. With the raised awareness of copyright issues this will probably become more common.

At present the Me framework is in the conceptual stage, with selected parts being prototyped. MVML based MMOs have been created and are being used as the source images to a random image generator on the front page of a research web site. A sniplet based content delivery system is being used with tertiary students that covers several domains. A business card application, where the cards are described with an image and associated MVML file is to be developed to track personal contacts. The construction of a "Me" frame work based prototype is in the design phase, with implementation in the near future.

References

1. DLNET: Brief Introduction to Learning Objects in DLNET (15-Jun-2002). Retrieved June 15, 2002, from http://www.dlnet.vt.edu/working_docs/reports/ARI_LO_Def.pdf
2. Davidson, C. & Voss, P. (2002) Knowledge Management. An introduction to creating competitive advantage from intellectual capital, Tandem Press. http://www.tandempress.co.nz. ISBN 1-877178-94-2
3. Dublin Core (http://dublincore.org/)
4. EndNote (n.d.). Retrieved August 04, 2004, from http://www.endnote.com
5. Guarino, N and Welty, C. (2002) Evaluating Ontological Decisions with OntoClean. *Communications of the ACM*, 45, 2. pp. 61-65.
6. Heitfield, E. (1996, Oct 10) Using BibTex and epsfig. Retrieved May 26, 2003, from http://emlab.berkeley.edu/wp/erik_bib.pdf
7. IEEE Learning Technology Standards Committee (LTSC) (1999) Learning Object Metadata, retrieved July 10,2003 from http://ltsc.ieee.org/doc/wg12/LOM3.6.html.
8. McGreal, R. and Roberts, T. (2001) A Primer on Metadata standards: From Dublin Core to IEEE LOM. Retrieved November 14, 2001, from http://vu.cs.sfu.ca/vu/tlnce/cgi-bin/VG/VF_dspmsg.cgi?ci=130&mi=1
9. Natu,S. & Mendonca, J. (2003) Digital Asset Management Using A Native XML Database Implementation *Proceeding of the 4th conference on Information technology education* , CITC4'03, October 16–18, 2003, Lafayette, Indiana, USA. 237-241
10. Pytches-Walker, J. (2003) Cyber Business Center, Glossary. Retrieved October 06, 2004, from http://www.nottingham.ac.uk/cyber/fullglos.html
11. Verhaart, M. (July 2000) Designing web pages for producing electronic and paper based teaching material! , In S. Mann (Ed) *Proceedings of the 13th Annual Conference of NACCQ*, Wellington.399-410
12. Verhaart, M. (2002, Dec) Knowledge Capture at Source. Developing collaborative shared resources. In Kinshuk, R. Lewis, K.Akahori, R. Kemp, T. Okamoto, L. Henderson, C.-H.Lee (Eds.) *International Conference on Computers in Education*, Auckland New Zealand., IEEE Computer Society, 1484-1485
13. Verhaart, M. and Kinshuk, Dr. (2003, June) An Extensible Content Management System using Database and Internet Technology, In D. Lassner, C. McNaught (Eds.), *Proceedings of ED-MEDIA 2003*, June 23-28; Honolulu, Hawaii, USA, AACE, 162-166
14. Verhaart, M. (2004) Learning Object Repositories: How useful are they? In S. Mann, T.Clear (Eds.) *Proceedings of the 17th Annual Conference of the National Advisory Committee on Computing Qualifications Conference*, Christchurch, New Zealand., July 6-9.465-469.
15. Verhaart M., Jamieson J. & Kinshuk (2004). Collecting, organizing and managing non-contextualised data, by using MVML to develop a Human-Computer Interface. In M. Masoodian, S. Jones, & B. Rogers (Eds.) *Lecture Notes in Computer Science, 3101*, 511-520.

16. What is a Digital Asset? (2003), BaseLine . Retrieved 2004, August 2 from http://common.ziffdavisinternet.com/download/0/1895/p057.pdf

9. A Didactics Aware Approach to Knowledge Transfer in Web-based Education

Denis Helic, Hermann Maurer, and Nick Scerbakov

Institute for Information Systems and Computer Media
Graz University of Technology
Inffeldgasse 16c, 8010 Graz, Austria
{dhelic,hmaurer,nsherbak}@iicm.edu

In this chapter we argue that the current Web-based educational systems, as well as the recent standardization efforts in this field are strongly technology-centric. Because of this, such endeavors usually neglect the didactic component of Web-based education thus becoming completely didactic-neutral.

However, recent research in the field shows that in order to achieve improvements in efficient knowledge transfer, learning outcome, or users' satisfaction didactics plays a crucial role. For example, such promising didactic approaches as collaboration, project-oriented learning, or experiential-learning need to be addressed by Web-based education systems and standards.

In this chapter we present an innovative system called WBT-Master that was built by our institute at Graz University of Technology. In WBT-Master technology is applied only as a vehicle that supports didactic aspects of Web-based education.

The main idea of WBT-Master is in the use of both conventional and innovative tools compatible with the current Web and Web-based educational standards to support didactics in a Web-based environment and in that way facilitate, more efficiently, transfer of knowledge from people who posses that knowledge to people who need to acquire it.

9.1 Introduction

It is our experience that most of the system development and standardization efforts in Web-based Education (WBE) field concentrate on the technological aspects of the field. Usually, these efforts neglect to a great extent the didactic side of WBE.

For example, the major commercial WBE systems, such as WebCT [30] or Blackboard [4] are mainly concerned with providing the tools to enable easy development, access, and work with Web-based educational material. From the didactic point of view such systems are based on learners taking a part in what can be simply called Web-based reading of educational material.

D. Helic, H. Maurer, and N. Scerbakov: *A Didactics Aware Approach to Knowledge Transfer in Web-based Education*, StudFuzz **178**, 233–260 (2005)
www.springerlink.com

Similarly, the standardization efforts such as Sharable Content Object Reference Model (SCORM) [2] developed by the international standardization bodies deal mainly with interoperability issues in WBE at the content level. Thus, such standards focus on creation, distribution, exchange and reuse of educational material in different WBE systems. However, SCORM compliant educational material usually does not carry any additional didactic-related information. Rather it only defines a simple navigational structure which prescribes how to navigate through the content. Such a navigational structure supports yet again only the most simple Web-based consumption of the educational content - Web-based reading.

However, we believe that WBE should support a wide range of different didactic approaches to enable efficient knowledge transfer in a Web-based environment. Here Web-based reading might be seen as just one (and for that matter the simplest one) of such didactic approaches. In such an approach to WBE the Web-based technology is applied only as a vehicle that supports didactic aspects of WBE, and ceases being the central part of it. Thus, the key here is in the use of both conventional and innovative tools compatible with the current Web and WBE standards to incorporate didactics into WBE and in that way facilitate, more efficiently, transfer of knowledge from people who posses that knowledge to people who need to acquire it.

We believe that the simplest way of introducing didactic aspects to WBE is by implementing tool support for a number of teaching and learning scenarios into WBE systems. For example, such typical scenarios as project-based learning, goal-oriented learning, or experiential-learning should be supported by tools and systems to cover a wide range of day-to-day training situations in both academic and corporate environment.

This chapter describes our efforts to implement this support into the WBT-Master, an innovative WBE system that was developed at our institute. The system supports a number of typical, as well as innovative teaching and learning scenarios applicable in WBE.

The remainder of this chapter is organized as follows. The next section describes in more details the current trends in WBE system development, research, and standards clearly identifying the above mentioned technology-centric problem of WBE. The third section provides an introduction to WBT-Master and lists teaching and learning scenarios supported by the system. The consequent sections describe in more details the listed scenarios at both levels - the didactic and the technological level. Finally, we present the results of an evaluation of the WBT-Master system and its concepts. The evaluation was conducted with hundreds of users in both academic and corporate environment and concentrated on estimating the improvements in learning outcome by using the traditional classroom approaches, the current WBE approaches, and the innovative approaches presented in the chapter. Moreover, issues such as learning experience, user satisfaction, usability, and others were addressed by the evaluation.

9.2 Current Trends in WBE

Recent surveys on the number of installations and distribution of WBE systems at universities in Europe [23], Australia [5] and the USA [6] show that the two most popular and used systems are WebCT and Blackboard. Approximately, the installations of these two systems constitute more than 60% of all WBE system installations.

Both of these systems can be categorized as so-called course management systems, i.e., they utilize the well-known "online course model". Usually, all the tools in such systems are optimized to support easy creation, publishing, update, and access to Web-based courses. Such Web-based courses are composed of a number of Web pages interrelated by means of computer-navigable links. Thus, the authoring tools allow authors to upload Web pages created on their local sites, or to create new Web pages directly with the system via integrated Web page editors. At the next step, the pages are related to each other and the publishing procedure makes the finished course accessible for learners. Additionally, the published courses are associated with a discussion forum which provides the communicational context for the learning sessions [21].

From the didactic point of view the facilities offered by these systems leave much to be desired. Usually, these systems support only traditional forms of teaching (e.g., a teacher prepares educational material that learners need to read) and encourage poor instructional design models. Consequently, the learning outcome and the users' satisfaction (of both teachers and learners) is quite low [20]. For example, a study conducted at the College of Nursing, University of Oklahoma, compared the learning outcome in a traditional vs. Web-based baccalaureate nursing research course. The study showed the following disappointing results. Overall, there was no significant difference in examination scores between the two groups as far as the three multiple-choice examinations and the course grades are concerned. Students who reported that they were self-directed and had the ability to maintain their own pace and avoid procrastination were most suited to Web-based courses [16]. Clearly, these results need to be improved tremendously.

The main reason for this situation lies in the fact that such systems are strongly technology-centric. Basically, the current WBE systems try to solve technology-based problems of WBE, such as authoring of educational material, reuse of Web-based courses, user management, efficient learners' progress tracking, and so on. Thus, the role of didactics is strongly neglected in such systems. For example, Mioduser refers to this situation as "one step ahead for the technology, two steps back for the pedagogy". However, the key must be in research and development of novel Web-based educational models and in the support of the current pedagogical approaches (e.g., use of inquiry-based activities, application of constructivist learning principles, and use of alternative evaluation methods) [19].

Similarly, other research work suggests development of pedagogy-aware Web-courses emphasizing that the strategies that enhance learning in the traditional classroom should be replicated in Web-based learning sessions. For example, these strategies include but are not limited to accommodating diverse learning styles, incorporating a good study guide with a content section, providing a communicative network and establishing a review process [10]. Clearly, the current WBE systems are far away from supporting such methodologies.

On the other hand, organizations such as IEEE Learning Technology Standards Committee, IMS Global Learning Consortium, and Advanced Distributed Learning Initiative (ADL) are working on standardization efforts in WBE. These efforts include the following standards:

- Learning Object Metadata (LOM) standard [15] for describing educational material with standardized metadata. This metadata should support learners and teachers in retrieving relevant educational material in an easy and efficient way. Further, it should provide the technical background for interoperability of educational material from different WBE systems. Thus, reuse and exchange of educational material between two different WBE systems is encouraged by this standard.
- SCORM suite of standards, such as SCORM packaging or SCORM Simple Sequencing [1]. These standards address issues of packaging Web-courses so that interoperability, reuse, and exchange of Web-based courses between different WBE systems is guaranteed. For example, SCORM packaging provides a standardized way of organizing the educational content into items, small packages, or even complex structures, and prescribes how such content can be navigated. SCORM Simple Sequencing goes one step further by specifying so-called learning paths, which can branch according to the current learning situation.
- IMS standards [11], such as IMS Metadata, IMS Content Packaging or IMS Simple Sequencing, which address similar issues as LOM or SCORM standards. Actually, many of the IMS standards, such as IMS Simple Sequencing are included in other standards such as SCORM Simple Sequencing.

Although such standardization efforts have many advantages, such as sharing and reuse of educational material, standardized way of packaging of educational material, flexibility in content presentation, interoperability across systems, they also have some disadvantages. The basic disadvantage is the total lack of addressing didactics and pedagogy in WBE. For example, SCORM claims to be "pedagogically neutral", which means that it is impossible to create SCORM packages that relate to some didactic approach, say project-oriented, or problem-solving learning approach [12]. In other words SCORM packages can not capture didactic relations between their components.

However, as recent research studies show incorporating didactics into WBE leads usually to far better results in learning outcome, learners' and teachers' satisfaction, learners' community building, and so on. For example, Hirumi shows in his study [9] that careful planning and design of interactions in WBE, where learners are supervised by means of immediate feedback, discussions and clear didactic and learning goals leads to improvement in learning outcome. Similarly, the study with deployment of project-oriented collaborative didactic approach conducted by King [14] shows tremendous improvements in building a community of learners, which helped to solve problems related to the project at hand. In the future, one of the keys to the learning process in WBE will be communication between learners themselves, learners and teachers, and the formation of learning communities held together by a common learning goal, which is modeled by a sound didactic approach.

9.3 WBT-Master: A Didactics Aware Approach to WBE

In WBT-Master didactic approaches are referred to as teaching or learning scenarios. These scenarios might be seen as a particular way of how different types (roles) of users work with the system, the system's tools and educational material available in the system to achieve a particular learning goal. Thus, each scenario can be described by:

- a particular way (i.e., a story) of how to achieve the learning goal,
- the user roles that are involved in the story,
- the system tools needed to support the story,
- educational material that is needed to achieve the learning goal.

For instance, the above mentioned Web-based reading scenario can be defined as the following teaching scenario:

- An author has a group of learners that need to improve their knowledge on a certain subject. Thus, the author prepares a number of Web pages containing relevant educational content and connects these pages with links in a navigable structure (i.e., course). After the course has been created the author publishes it in the system. Now, the learners access the published material and read through it to improve their knowledge about the subject. During their work with the published material the learners communicate with the author via the attached discussion forum. Additionally, the author tracks the progress of the learners by means of different statistic tools.
- The user roles involved in the scenario are authors and learners.
- The system tools needed to support the scenario are the authoring tool for preparing Web-based educational material (i.e., courses), and a standard Web browser to access and work with educational material.
- Educational material comes in the form of a number of Web pages, which contain relevant educational content. The pages are connected by links into a navigable structure.

With the teaching and learning scenarios, which are implemented in WBT-Master we tried to take into account recent advancements in the traditional classroom didactics, as well as in the technology enhanced didactics. These scenarios incorporate such promising didactic approaches as project-based learning, problem-solving, constructivist approaches, collaboration, and so on.

In this chapter we present the following teaching and learning scenarios from WBT-Master:

- Web-based reading - this basic WBE scenario was extended in WBT-Master by sophisticated communicational and collaborative features such as annotations.
- Web-based tutoring - a teaching scenario where a tutor works with a group of learners in both synchronous and asynchronous mode, leading them to achieve a particular learning goal.
- Knowledge profiling - a scenario supporting the acquisition, structuring, and reuse of extracted expert knowledge.
- Knowledge mining - a learning scenario where learners are supported in exploring extracted expert knowledge by means of personalized knowledge retrieval facilities.
- Project-oriented learning - a learning scenario where a group of learners works together on a project, e.g., a software engineering project.

In the remainder of the chapter each scenario is described according to the above introduced template, i.e., the user roles, the system tools, educational material and the story of the scenario are given. After the scenario was defined we firstly discuss didactic aspects of that approach. Then, the technical issues such as software requirements, technical problems and obstacles, as well as possible solutions are discussed. Finally, we present an example and a screenshot of a typical learning session with that particular scenario.

9.4 Web-Based Reading

The Web-based reading scenario was already defined in the previous section. The support for this simple teaching scenario in WBT-Master closely reflects the similar support in other WBE systems, such as WebCT or Blackboard. However, there are also a few significant differences.

Didactically, the Web-based reading scenario in WBT-Master was extended by the promising collaborative facilities, such as annotations and synchronous and asynchronous communication [17]. Thus, the learners and the author can add and change the content by annotating it for themselves or others. Other users can even annotate the notes previously made, in this way activating a powerful communication channel. Each annotation has a certain type, such as "Comment", "Question", "Answer", etc., which provides a communicational context that can be very important in the learning process [24].

Moreover, annotations can also take the form of links, i.e., material can be linked together by the learners for their own benefit and for the benefit of the whole group. Thus, learners themselves contribute to the content on-the-fly.

Further, the Web-based reading scenario in WBT-Master supports other ways of synchronous and asynchronous communication between the learners and the author. These communicational facilities include chat rooms and the attached discussion forum. Note, that all the annotations that were made previously within the context of the educational content are also accessible via the discussion forum.

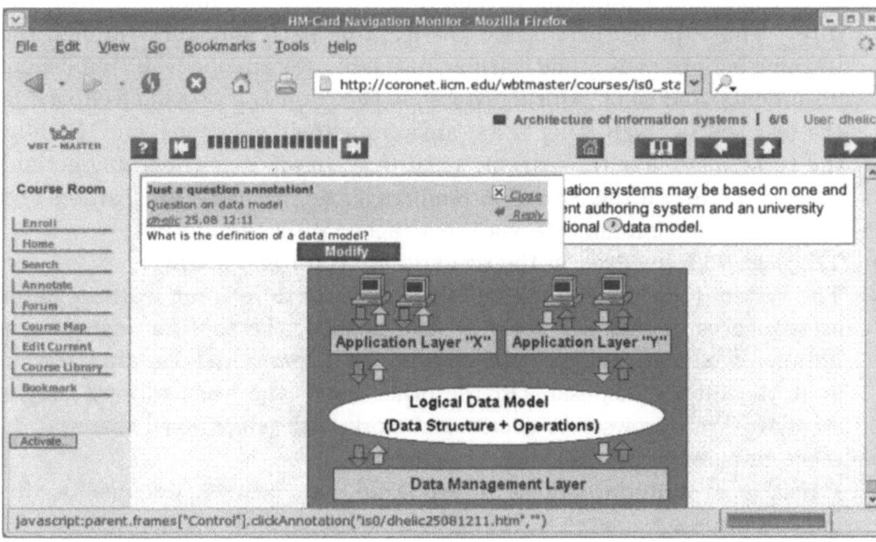

Fig. 1. Annotations in WBT-Master

Technically, educational material in the Web-based reading scenario is SCORM compliant. This ensures modularity, reuse, and interoperability of the educational content units. For example, suppose that we developed an educational unit about "Relational Data Model". Now, this unit might be reused in a number of different contexts. For instance, it can be reused in the context of another educational unit dealing with "Databases" in the practical sense, but it can also be reused in the educational unit dealing with the theoretical aspects of "Data Models".

9.5 Web-Based Tutoring

The Web-based tutoring scenario might be defined as the following learning scenario:

- A tutor has a group of learners that need to achieve a particular learning goal, e.g., the learners need to acquire some knowledge in a particular subject matter or they need to learn how to solve a particular kind of problems. The tutor defines a learning path, which the learners need to follow in order to achieve the learning goal. Such a learning path is a sequence of learning actions that need to be accomplished in a step-by-step manner by the learners. Each learning action comes with some educational material which should be consumed by the learners at that particular step. Alternatively, a learning action might be associated with a test that the learners need to pass, a request for publishing a document, a request to solve a particular problem, or simply a request to communicate with the tutor. Thus, the learners access the learning path and work through the learning actions consuming learning resources, working on tests, publishing documents, and so on. During this time the tutor can provide feedback to the learners by evaluating tests, answering their questions, etc.. Finally, the tutor may alter the learning path (e.g., insert a new learning action) as long as the learning situation requires it. Additionally, the learners and the tutor can communicate via the attached discussion forum.
- The user roles involved in the scenario are tutor and learners.
- The system tools include the authoring tool for developing learning paths as sequences of learning actions. Additionally, the tool for managing a library of educational material needed for a particular learning session is at the tutor's disposal. On the other hand, the learners need only a standard Web browser to access and work with educational material, to make tests, or to publish their documents.
- Educational material can be of any kind, i.e., courses, documents, discussion forums available in the system or external Web resources such as external Web pages.

The Web-based tutoring scenario reflects the well-known goal-oriented didactic approach [29]. Thus, the tutor leads the learners to achieve a particular learning goal. For example, a particular learning goal for software engineering students might be to learn how to write the user requirements document for a software system.

In the Web-based tutoring scenario the learning goal is achieved by following a predefined sequence of learning actions, i.e, the learning path. Since there are different types of learning actions, such as reading, writing, solving a problem, answering questions and others, the learning process can be based on sophisticated instructional models. For example, the above mentioned software engineering students, after having read more theoretical documents, might become involved in writing a sample user requirements document to gain practical experience. Note here the difference between the Web-based tutoring and the Web-based reading scenario. In the Web-based reading scenario the learners are supposed to reach their learning goal by simply following the navigational sequence and reading the educational content. Consequently, the Web-based reading scenario cannot prescribe a writing assignment for the software engineering students as the Web-based tutoring scenario can.

Another important aspect of the Web-based tutoring scenario is the immediate feedback by the tutor to the current learning situation. At each particular step of the learning session the tutor can provide feedback to the learners by communicating with them, evaluating their contributions, or alternating the learning path if new learning actions need to be inserted. For example, the tutor recognizes that the learners did not understand some concept entirely (e.g., by looking at the test result) and that they need some additional information. Thus, the tutor decides to insert a new learning action into the learning path attaching to it a document that provides the needed information. Note that the feedback can be provided for the whole group of learners, as well as for a single learner. In this way, the learning actions are customized to the current learning needs, learning situation, knowledge level, and learning preferences not only of the whole group but also of a particular learner. Thus, each learner's learning experience can be highly personalized by the tutor.

From the technical point of view, the tutor is involved in managing the library of educational material and in creating and manipulating the learning path. Note that the tutor is supposed to reuse all educational material available in the system by including it into the library. This material is then being referenced from within the learning path. To ensure interoperability between different scenarios and to enable reuse we decided again to apply the SCORM standard for defining the educational content (i.e., packaging of the content) and the learning path. The SCORM standard includes the simple sequencing model, which provides means for defining learning paths and rules for choosing between different alternatives. Although the simple sequencing model cannot capture all aspects of the Web-based tutoring scenario (e.g., altering of the learning path based on communication between the tutor and the learners) it can be seen as a solid basis for further development. Note also that our solution works as an authoring tool for such simple sequence models because the sequences are defined by the tutor on-the-fly taking into account the current learning situation. This can be seen as the additional value of the tool because authoring of such sequences before any learning session starts is usually a very hard task.

9.6 Knowledge Profiling

The knowledge profiling scenario is defined as the following learning scenario:

- A teacher (e.g., an author or a tutor) has a group of learners that need to improve their knowledge on a certain subject. The organization with which the teacher and the learners are affiliated manages a large repository of extracted expert knowledge in the form of documents, external Web pages, lessons learned, discussion forums, etc. To ensure that the learners

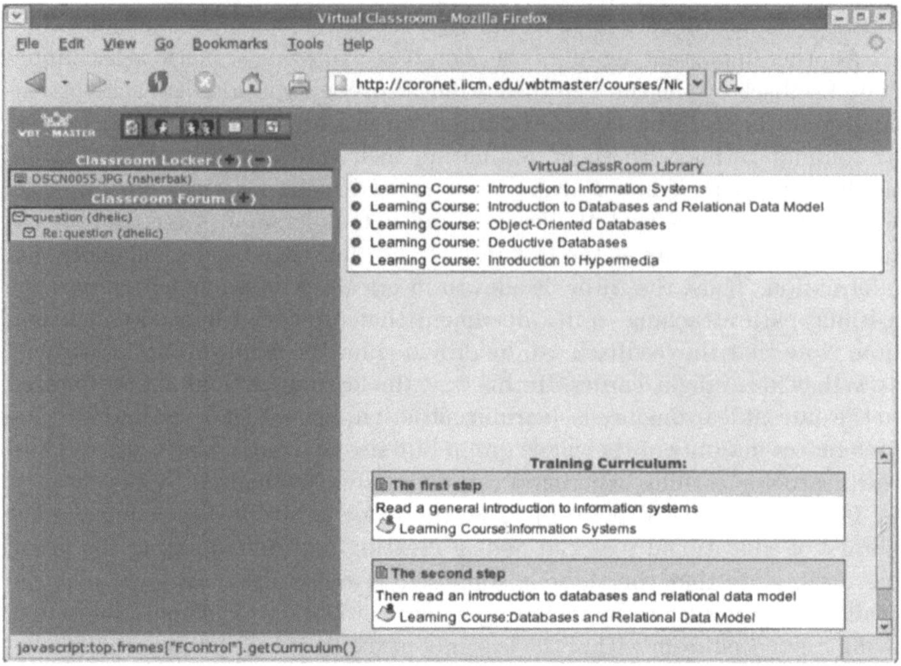

Fig. 2. Working with Learning Actions

learn from the expert knowledge the author develops a domain ontology of the subject and classifies the expert knowledge according to the ontology. Finally, the learners access and work with the classified knowledge.

- The user roles involved in the scenario are teachers (either authors or tutors, or both), learners, and indirectly experts.
- The system tools which are needed include the authoring tool for developing a domain ontology, the authoring tool for classifying extracted expert knowledge according to the developed domain ontology, and a standard Web browser for learners to access and work with the classified expert knowledge.
- Educational material in this scenario is extracted expert knowledge coming in the form of internal documents, external Web pages, discussion forums, etc.

From the didactic point of view, this scenario addresses a few issues of the so-called experiential-learning. The experiential-learning is related to learning with experiences [7, 8]. The main concern of the experiential learning is how to transfer these experiences efficiently. Such experiences are mostly provided by experts, where the expert experience results from many years of practice. The problems in the experiential learning come from the fact that the knowledge of experts is somehow routine and difficult to make explicit or understandable to novices. Some of the reasons for this situation are the lack of the background knowledge of the domain (the declarative knowledge) and the lack of anchoring between experiences and the declarative knowledge [18]. The knowledge profiling scenario tries to bridge this gap in the experiential-learning by providing a high-level declarative description of the domain (i.e., the domain ontology), and by linking the expert experiences to the ontology, i.e., by classifying the experiences to the categories and the relations of the domain ontology.

For example, suppose we have the extracted expert knowledge in the domain of databases and information systems. This knowledge is comprised of, let say, a single general document about information systems and a single "lessons learned" document on the practical development of database systems. The domain ontology might include two categories: the "Information Systems" and the "Databases" category. Since database systems are a special kind of information systems the domain ontology might relate the "Information Systems" category to the "Databases" category by means of the "includes" relation. Additionally, the ontology might include the inverse relation of the "includes" relation, i.e., the "isKindOf" relation. Finally, we might classify the general document about information systems to the category "Information Systems" and the "lessons learned" document to the "Databases" category. Note, that the two documents are now automatically anchored within the background knowledge, i.e., they are explicitly related by means of the "includes" and its inverse "isKindOf" relation.

Technically, the implementation of the knowledge profiling scenario needs to meet the following requirements. Firstly, the system must support the development of domain ontologies or seamless integration of domain ontologies developed with external ontology editors. To ensure interoperability between the system and external tools domain ontologies should be developed by means of standardized knowledge representation techniques, such as recently developed RDF Schema [32] and OWL [31] languages.

Secondly, the system needs to support the teacher during the classification of the extracted expert knowledge by means of automatic and semi-automatic methods. For example, the system might suggest to the teacher that a particular document should be included into a specific domain category. There are a few different ways to support automatic or semi-automatic document classification, such as metadata management or full-text processing. In WBT-Master we apply metadata for this purpose, since implementing document classification is usually very hard in a WBE environment. Usually, such an environment deals with heterogeneous documents (e.g., Web pages, discussion forums, internal documents in different formats, etc.), which makes supporting of full-text processing very difficult.

Also, metadata management and especially metadata gathering in such an environment can be very expensive since the users of the system need to provide metadata manually. In WBT-Master we apply a semi-automatic approach for metadata gathering. Thus, the system manages sophisticated user profiles, which contain information of users' field of expertise, users' general interests, users' current involvement in the learning and teaching processes, and so on. Then, the system tries to apply this information to automatic generation of metadata. For example, suppose we have an expert in "Databases". The expert declares the "Database" expertise in his/her user profile. Now, whenever this expert contributes a document to the system, the system automatically adds a metadata description to the document stating explicitly that this document deals with "Databases". Then, this information can be used during the classification process.

Finally, the system needs to support learners in their work with the domain ontology and the classified expert knowledge, i.e., the categories of the domain should be searchable and navigable. For example, the learners might want to search for all documents belonging to the "Databases" category, or they can navigate through the domain ontology and in that way reach the "Databases" category and its documents (e.g., by following the link "includes" emanating from the "Information Systems" category).

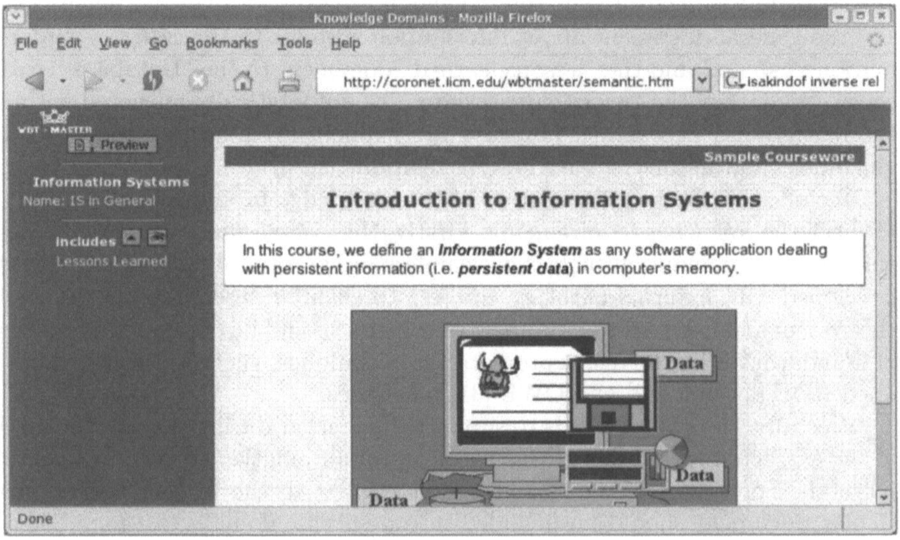

Fig. 3. Accessing Documents Anchored to Background Knowledge

9.7 Knowledge Mining

The knowledge mining scenario might be seen as a refinement of the knowledge profiling scenario. Thus, this scenario is built up on the same infrastructure, i.e., the domain ontologies and the classification of the extracted expert knowledge by means of these ontologies. The main difference is in the way how learners access that knowledge. In the knowledge profiling scenario learners had to navigate or search through the ontology categories to find the relevant information. Taking into account that a typical domain ontology can include hundreds, even thousands, of categories and relations this might be seen as a rather tedious task. Moreover, in a typical training situation in a corporate environment learners need to find relevant information easily and quickly, and usually do not have time to navigate through thousands of categories. In such a common on-demand training situation, the knowledge mining scenario tries to provide support for learners in their initial access to relevant information.

Didactically, the knowledge mining scenario addresses yet another issue related to the experiential-learning. This issue deals with the way how learners access relevant information, i.e., how they find such information in an efficient manner.

From the technical point of view, the knowledge mining scenario extends the knowledge profiling scenario in the following way. In the knowledge profiling scenario we were mainly concerned with creation of the domain ontologies and the classification of the expert knowledge to the categories from the ontologies. One of the methods applied to automatic or semi-automatic execution of this process was a management of user profiles, and description of users' field of expertise with metadata. In the knowledge mining scenario we extend the notion of user profiles by describing what are the fields of interest of each particular learner. This information is then used to facilitate the initial access to relevant information.

For example, suppose a learner is interested in information systems. Now, whenever the learner accesses the categorized expert knowledge, say by navigation, the system provides links to all documents containing expert knowledge on the topic of information systems. Similarly, search mechanism profits from the same information by ranking documents dealing with information systems at the top of search results.

Finally, the system makes use of the another important property of domain ontologies, i.e., inference. Inference is a technique that supports deduction of new facts, e.g., automatic classification by investigating categories, relations and their properties in domain ontologies. Recollect the example that we introduced above, i.e., we have two categories: the "Information Systems" category and the "Databases" category. These two categories are related by means of the "isKindOf" relation, i.e., the "Databases" category "isKindOf" the "Information Systems" category. Usually, the "isKindOf" relation is defined as transitive, i.e., if "A" "isKindOf" "B", and "B" "isKindOf" "C", then "A" "isKindOf" "C", thus allowing the principle of subsumption to be applied. Obviously, if a document is classified to the "Database" category then it can be (because of transitivity of the "isKindOf" relation) automatically classified to the "Information Systems" category. Now, whenever the learner who is interested in information systems access that category the system automatically provides the learner not only with links to the documents in information systems, but also with links to the documents in databases.

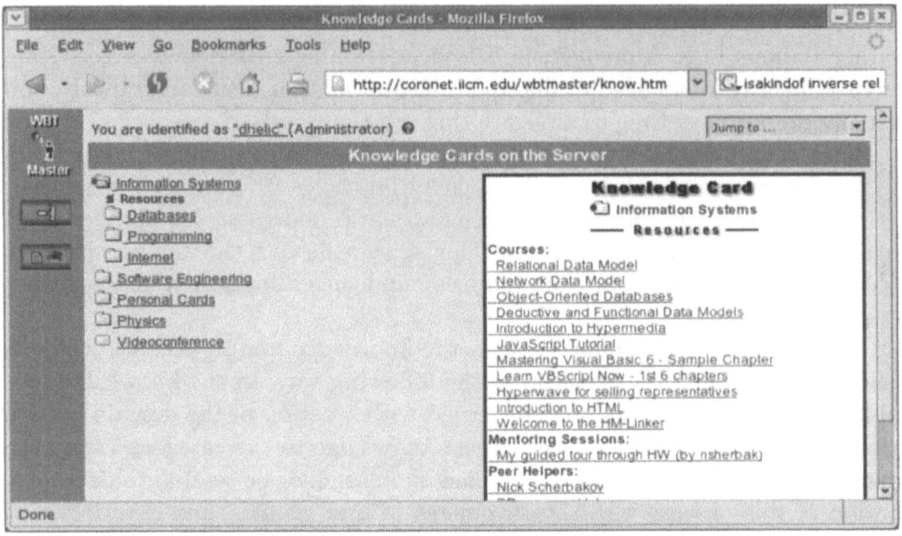

Fig. 4. Retrieving Documents Anchored to a Specific Category

9.8 Project-Oriented Learning

The project-oriented learning scenario can be defined as the following learning scenario:

- A teacher (an expert, an author or a tutor) has a group of learners that need to gain a practical experience in project-based collaborative work, e.g., working in a group on a software project. The teacher initiates a project-based learning session by creating a detailed project plan with the project steps and the time plan. At the next step, the teacher provides a sample project, which shows all the steps of a successfully executed project. Finally, the teacher provides a number of project alternatives for the learners. The learners constitute a number of teams, where each team selects one of the possible project alternatives as their practical example. The system provides communication tools, such as a discussion forum, a chat room, as well as collaborative facilities, such as version control system, annotations, tools to write project documents in collaboration, and so on. The teacher monitors the progress of the learners and provides feedback when necessary.

- The user roles involved in the scenario are teachers and learners.
- The required system tool is the integrated project management room, which provides facilities for creating and managing project plans, sample projects, project alternatives, as well as communicational and collaborative facilities.
- Educational material in this scenario consists of the project plan, the sample project, as well as external resources that may be linked to the project room via the discussion forum or annotation facilities.

Didactically, project-based learning is a model of learning that organizes learning around projects. Usually, projects are complex tasks, based on challenging questions or problems that involve learners in design, problem-solving, decision making, or investigative activities, furthermore they give learners opportunity to work relatively autonomously over extended periods of time; and culminate in realistic products or presentations [13]. Other defining features found of project-based learning paradigm include authentic content, authentic assessment, teacher facilitation but not direction, explicit educational goals, cooperative learning, reflection, and incorporation of adult skills .

Crucial for a successful and effective application of such a project-based learning paradigm is the careful developing and planning of effective projects. The basic properties of such effective projects might be summarized as follows [27]:

- Learners should be put at the center of the learning process.
- The project work is central to the curriculum.
- The project must motivate learners to explore important topics on their own.
- Project management should be accomplished by using appropriate tools, such as computer-based project management tools.
- The project outcome or the result that learners need to produce must include learning techniques such as problem solving, in-depth investigation of topics, research, reasoning, and so on.
- The project should include outcome alternatives that learners might choose from, or that they can work on one after another applying the experience they gained before.
- The project must be collaborative, that is learners might work together in small groups, or they can present and discuss their partial and complete results with other learners at any time.

Let us look now at an example of a project-based learning course. In a study reported by Barron [3], learners worked for five weeks on a combination of problem-solving and project-based learning activities focused on teaching learners how basic principles of geometry relate to architecture and design. The problem-solving component involved helping to design a playground in a simulated computer aided environment. The project-based component involved designing a playhouse that would be built for a local community center.

Following experience with the simulated problem, learners were asked to create two- and three-dimensional representations of a playhouse of their own design and then to explain its features in a public presentation to an audience of experts.

Recently, numerous research papers on project-based learning have been published showing the benefits of this learning paradigm for learners and teachers as well. For example, these reports show tremendous gains in learner achievements, large gains in learners' problem solving capabilities, gains in learners' understanding of the subject matter, perceived changes in group problem solving, work habits, and other project-based learning process behaviors [26].

Technically, the project-based learning must be supported by means of an integrated project management tool. We implemented such a project-management tool in WBT-Master. This tool consists of the following components:

- A special document (curriculum) describing in few words the course and project motivation, problems that need to be solved, goals, etc.
- A special discussion folder providing a sample project with the definition of project plan, i.e., number of project steps and the time table for these steps. Each step is documented with a number of publications.
- A number of project discussion folders, which provide project alternatives for learners to chose from. These folders hold also all learner contributions.
- A number of collaboration and communication tools, such as online presence lists, chat rooms, discussion forums, etc.
- Evaluation tool for teachers evaluating learners' work.

9.9 Evaluation of WBT-Master Concepts

WBT-Master was developed within the scope of CORONET (Corporate Software Engineering Knowledge Networks for Improved Training of the Work Force) project funded by the European Union. The CORONET project was running from Mai 2000 until Mai 2002. The project consortium consisted of:

- Center for Advanced Empirical Software Research, the University of New South Wales, Sydney, Australia
- Atlante, Madrid, Spain
- DaimlerChrysler, Ulm, Germany
- Fraunhofer Institute for Experimental Software Engineering (IESE), Kaiserslautern, Germany
- Fraunhofer Institute for Computer Graphics (IGD), Darmstadt, Germany
- Highware, Paris, France
- Institute for Information Systems and Computer Media (IICM), Graz University of Technology, Austria

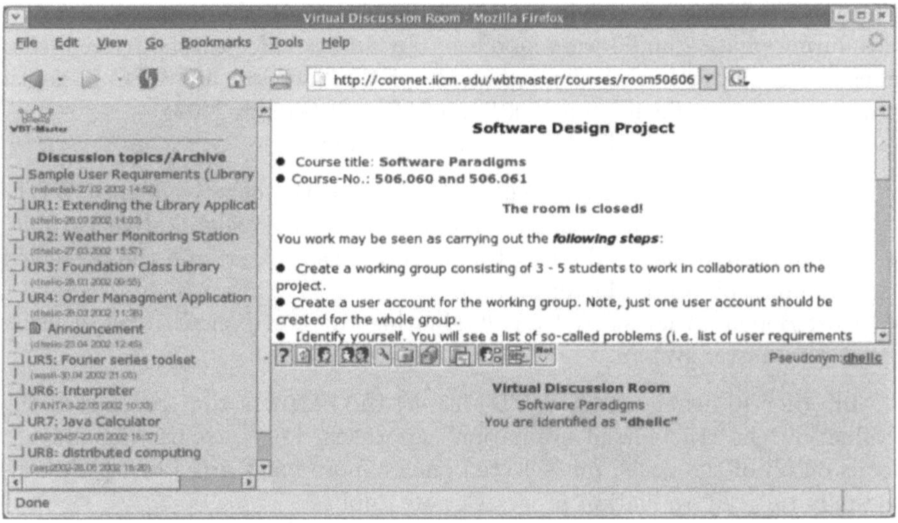

Fig. 5. Project Management Tool in WBT-Master

- Centro de Computacao Grafica, Coimbra, Portugal

WBT-Master was mainly developed by the IICM. The application partners in the project were DaimlerChrysler, both Fraunhofer institutes, and Highware. These institutions deployed WBT-Master and evaluated it in a wide range of possible applications. Additionally, we at the IICM used the system for hundreds of university students during lectures at our university.

9.9.1 CORONET Project Evaluation Approach

The CORONET project evaluation activities were performed through the following 3 phases [28]:

- Phase 1 (June - August 2001): In-depth assessment of the first WBT-Master prototype. The results from this evaluation were the main input for the enhancement of the CORONET methodology and infrastructure during the 2nd cycle of the CORONET project.
- Phase 2 (September 2001 - April 2002): Continuous evaluation studies performed with the WBT-Master prototype in parallel with incremental enhancements of the product.
- Phase 3 (February - April 2002): In-depth assessment of the improved WBT-Master prototype.

The three phases were performed in a systematic way during the project according to a detailed evaluation plan developed during cycle 1 of the CORONET project. The evaluation approach was finely tuned with the contribution of software and learning evaluation experts involved as members of the

CORONET Pedagogic Advisory Board. Furthermore, the evaluation activities during phase 2 and 3 were monitored by additional requests derived from phase 1 results and from the comments of the 2nd and 3rd CEC in-depth project reviews in Paris (June 2001) and Madrid (March 2002).

General goals for WBT-Master evaluation were:

- Analysis of learning effectiveness: Evaluating the effectiveness of WBT-Master system in supporting knowledge sharing and collaborative learning.
- Usability analysis: Evaluating the perceived ease of use and the perceived usefulness of WBT-Master system.
- Cost-Benefit Analysis (CBA): Evaluating the cost-benefit ratio of using WBT-Master system.

In order to evaluate the WBT-Master, the partners that conducted the evaluation chose a "mixed evaluation" approach, i.e., each partner did not focus on all of the goals, but selected one or more focus areas to which individually tailored evaluation processes were applied.

To analyze learning effectiveness DaimlerChrysler, consistent with its role as a software development organization, focused on evaluating the effectiveness of WBT-Master in supporting continuing, self-directed, collaborative learning. The evaluation process was based upon the cognitive load theory and relied on a series of specifically designed evaluation sessions that were conducted in a specifically established evaluation laboratory setting, involving members of the research group as well as members of a business unit. Fraunhofer, consistent with its role as a research institute, focused on evaluating the effectiveness of WBT-Master in supporting collaboration and knowledge sharing. This was done by conducting two quasi-experiments that compared the efficiency and effectiveness of conducting similar tasks with and without using WBT-Master. Highware, consistent with its role as a training service provider, focused on evaluating the effectiveness of WBT-Master in supporting web-based learning by training and web-based experience sharing. Evaluation data was collected with the help of specifically designed questionnaires.

To perform usability analysis Perceived Ease of Use (PEU) and Perceived Usefulness (PU) of WBT-Master from the point of view of end users was estimated. In order to analyze the PEU and PU end users were requested to answer related sets of questions. For data collection, specifically designed questionnaires were distributed to end users of WBT-Master at the end of the trial period for a particular learning scenario. In order to test the reliability of their analysis results Fraunhofer IESE calculated and interpreted Cohen's Kappa coefficient. DaimlerChrysler used the questionnaire ISONORM 9241/10, which was evaluated with respect to validity and reliability. The ISONORM questionnaire was derived from the software ergonomic standard DIN EN ISO 9241.

Finally, Fraunhofer IESE designed and guided the cost-benefit analysis. Cost-benefit data was collected by DaimlerChrysler and Highware with the help of specifically designed data collection forms. The cost-benefit analysis was based on a 3-phase learning reference model. The purpose of this model is to provide a common basis for comparing different learning and training approaches in one common framework. The reference model consists of the following main phases:

- Pre-Learning Phase: comprising all relevant activities before learning is performed.
- Learning Phase: comprising all relevant activities during learning.
- Post-Learning Phase: comprising all activities taking place after learning is finished.

For each of the phases, during the evaluation studies conducted by the application partners DaimlerChrysler and Highware, associated cost and benefit data was collected.

9.9.2 Evaluation Results in Corporate Environment

The first evaluation goal focused on the effectiveness of learning with WBT-Master from the perspective of software organizations, software engineering research organization, and software training service providers, i.e., DaimlerChrysler, Fraunhofer IESE, and Highware. In accordance with the findings related to the first goal, all partners appreciated the innovative concepts offered by WBT-Master.

Nevertheless, the results of the evaluation studies related to learning effectiveness were not fully consistent. The data reported by scientists, software engineers, and software trainers at Fraunhofer IESE, Highware, and Highware's partner and customer organizations generally indicated improved learning effectiveness when using WBT-Master. The analysis of learning effectiveness conducted by DaimlerChrysler was partly influenced by negative judgment of the usability of the WBT-Master platform. This was reflected by the data received from DaimlerChrysler system users who expressed the feeling that the cognitive load associated with tool usage prevented them from learning in the proper sense.

In addition to the evaluation studies conducted by DaimlerChrysler, Fraunhofer IESE and Highware within the scope of the CORONET project, a large number of students (more than 100) at the Graz University of Technology have been using WBT-Master extensively since mid-2001 without major problems, thus confirming that the system can be considered a helpful instrument for collaborative learning and knowledge sharing, see sect. 9.9.3.

As a by-product of the analysis of learning effectiveness, some observations and conclusions on cultural and organizational aspects could be drawn from the associated evaluation studies. The analysis of WBT-Master user profiles clearly showed that there was a positive predisposition to work with a web-based learning environment as most of the users had been familiar with ICT for more than two years. However, some cultural factors were detected as being critical. They should be taken into account when introducing and operating WBT-Master.

First, shifting to e-learning clearly requests changes in the behavior of nearly all the participants involved. The changes are mainly related to:

- Learning approach: shifting from the conventional presence learning mode to using the Internet is not obvious for learners who have not yet had experience with or have not been prepared for using the new learning and knowledge transfer processes offered by a web-based learning environment.
- Pedagogical approach: replacing interpersonal relations which typically occur in conventional classroom settings by interactions between the learner and the web-based learning environment requires new competence on the part of trainers, tutors, and authors of learning materials.

Using a learning environment like WBT-Master is not a one-shot experience: it is highly recommended to properly introduce both the methodology and the infrastructure to all types of users in order to facilitate the adequate use of the learning environment. It clearly appeared from all evaluation studies that system users need some time to handle the new environment before focusing on any specific learning activity.

Here are some highlights from the evaluation at DaimlerChrysler. In total, forty individuals were involved in DaimlerChrysler's evaluation studies. Twelve of them actively participated in thirty-four in-depth evaluation sessions. The qualitative analysis applied to the "think-aloud protocols" and recorded video tapes of the evaluation sessions indicated that the concepts offered by WBT-Master (e.g., to combine collaboration and document work) were generally appreciated by system users. The following functionalities were considered most beneficial for the specific setting of DaimlerChrysler's evaluation study:

- Inference enabled ontologies for self-paced worker's knowledge mining,
- Web-based tutoring for the experts to give advice, and
- Various collaboration tools, i.e., forums, to collaborative problem-solving with peer learners, and collaborative knowledge building.

The positive impression, however, was negatively influenced by the subjective perception that the current version of the learning prototype platform WBT-Master was too difficult to use. In particular, the various options for communication/collaboration were perceived as too numerous, too spread out, and too hard to differentiate. One possible explanation for these partly negative results is that DaimlerChrysler's software engineers have to cope with extremely high pressure to continuously upgrade their knowledge on-the-job, possibly without being able to spend any effort other than the project-related effort on learning. Hence, this highly specialized clientele is used to (and needs to) work with a software environment that perfectly matches their specific expectations and does not require any introduction and learning curve. Thus, the tolerance level regarding expectations and actual behavior of a new learning environment is rather low. This might explain why WBT-Master, having the maturity of a research prototype, had problems to meet the high expectations of DaimlerChrysler's trial users.

Some highlights of the Fraunhofer IESE's evaluation are as follows. In total, seven individuals actively participated in Fraunhofer IESE's evaluation studies. The results of the evaluation studies show that knowledge sharing activities in a research department setting can be performed more efficiently and more effectively with using the knowledge transfer functionality offered by WBT-Master.

Finally, some of the results of the evaluation at Highware are as follows. In total, thirty-two individuals actively participated in Highware's evaluation studies. The results of the evaluation studies were in its majority positive. The main findings can be summarized as follows:

- The concepts contained in the learning methodology of WBT-Master are presented in a clear and concise style so that learners, trainers, tutors, and authors can easily identify the right learning scenario for their particular learning/training needs. This was particularly true for the scenario Web-based tutoring, which was the main focus of Highware's evaluation studies.
- The Web-based tutoring functionality provided by WBT-Master offers a viable alternative to classical in-class training settings. The effectiveness of virtual classes was judged as being at least as effective as conventional in-class sessions.
- Regarding the effective support of web-based experience sharing, the second focus of Highware's evaluation studies, WBT-Master successfully helped to establish a network of geographically distributed learners. From the point of view of the management, the establishment of such a network, facilitating learning at the workplace by connecting people to a network of distributed learning resources (documents, courseware, peers and experts) was considered as one of the main strengths of the CORONET system.

Since the main focus of the CORONET project was to develop innovative solutions to support collaborative web-based learning, it was not surprising that evaluation results judged WBT-Master as being acceptable as a training management system, but several proposals for future enhancements were made.

It was interesting to observe that the results of the evaluation studies conducted within Highware and in collaboration with Highware's partner and customer organizations in real business cases were more positive than the results of DaimlerChrysler, the other industrial partner in the CORONET consortium. One possible explanation is the following: Since the use cases defined by Highware were more focused, relatively less effort for introducing WBT-Master to their own staff and to their customers was needed. This helped to avoid misunderstandings of the concepts and paved the way for better tool acceptance. It also meant that both Highware staff members (who used the system internally) and end users of Highware's partners and customers were more tolerant towards a prototype system that obviously is not yet perfect (and thus imposes an initial learning curve) but delivers innovative functionality. Another explanation might be the differences between organizational cultures involved in DaimlerChrysler's and Highware's evaluation studies. Both DaimlerChrysler and Highware (and its partners and customers) are highly professional and successful in their respective businesses, but the individuals involved in DaimlerChrysler's evaluation sessions mainly work in complex team-oriented organizational settings with a strong product focus, whereas the individuals involved in Highware's evaluation sessions mainly work in small to medium sized network-like organizational settings with a strong service focus. Whether cultural differences induced by the different geo-political settings of the studies could also account for the different findings was not investigated.

Case studies to analyze perceived ease of use (PEU) and perceived usefulness (PU) of WBT-Master were conducted by DaimlerChrysler, Fraunhofer IESE, Fraunhofer IGD, and Highware. The data analysis of the various evaluation studies did not result in a consistent view. While subjective user acceptance of WBT-Master by individuals involved in DaimlerChrysler's evaluation studies turned out to be insufficient, evaluation data provided by WBT-Master users at Fraunhofer IESE, Fraunhofer IGD, and Highware (including their partner and customer organizations) showed positive PEU and PU ratings.

An explanation for this inconsistent result could be the different approaches that were chosen to conduct the evaluation studies. DaimlerChrysler based their analysis on very complex use cases that require a relatively high usability of the tool environment in order to avoid that system users resign from trial experiments. Given that WBT-Master is a prototype platform - and not a fully developed product - time constraints coupled with high expectations of the system users and their low level of tolerance towards initial learning curves resulted in low usability ratings. Hence, the partial non-acceptance of the system by DaimlerChrysler users might be perceived as a confirmation of the project intention to develop a usable prototype, but not an "off-the-shelf" software product. Due to the different nature of their use cases and the associated learning scenarios it was less difficult for the other partners (Highware, Fraunhofer IESE and Fraunhofer IGD) to introduce the system as a working prototype to their respective end users. As a consequence, in their evaluation studies these partners better managed to focus on innovative functionality and to invest into providing additional learning and user support (which would not be needed for a software product). This led to the positive results of the usability studies of these partners, reconfirming the overall project success.

The results of the Cost-Benefit Analysis (CBA) were gained from data collected by DaimlerChrysler and Highware in seven evaluation studies conducted across two evaluation cycles. The majority of the results showed that using WBT-Master is - in addition to the non-monetary benefits generated by the innovative methodology and infrastructure - beneficial from the monetary perspective. The CBA showed that:

- For Highware, a training provider, using WBT-Master increases the Net Present Value (NPV) and thus can be considered as being monetarily beneficial.
- For DaimlerChrysler, which is not a training provider, using WBT-Master does not generate a positive NPV in a short term. However, using WBT-Master over a period of more than three years is expected to result in a positive monetary effect.

Generalizing the CBA results, it can be expected that:

- Training providers can be advised to buy and apply WBT-Master "as-is" because cost savings along with a profit increase caused by travel cost reduction, reuse of training materials, and additional (new) customer services (based on the CORONET features) that generate additional revenue can be expected.
- Customers of training providers will experience - besides the non-monetary benefits of CORONET - a cost reduction through reduced inter-location travel of the employees attending to the CORONET-based training.
- Non-training providers, i.e., software development organizations, can reach a positive NPV in the training and learning cost by using CORONET for a few years.

9.9.3 Evaluation Results in Academic Environment

We applied the project-oriented learning scenario to conduct the 2002 summer term course on Software Engineering at the Graz University of Technology with more than 200 students. The Software Engineering course at our university consists of:

- Lectures on basic software development paradigms and vocabularies applied to describe the development paradigms and development processes.
- Software development project where students develop a software application following one of the presented development methods.

Thus, the practical part of this course is already project-oriented. Consequently, we wanted to conduct this project by means of WBT-Master. Thus, we prepared a special project-oriented learning session for the Software Engineering project. The session included the following items:

- Curriculum for the project, where we described the learning goals, learning mode, presented time schedules, etc.
- A sample software development project clearly identifying the development method, development process, and all steps that students needed to accomplish to successfully finish their projects.
- Four software development proposals, from which students chose their own projects.

The integrated project management tool provided all necessary facilities needed to conduct a Web-based software development project, for both teachers and students. Thus, students made their accounts, groups, and assigned their accounts to the groups. They posted their results as multimedia replies to a particular project folder, following the steps from the sample project. Communicational tools were available for them at any particular time. Teachers were able to track students' progress, evaluate the students' results and provide them with valuable comments. Discussion forum was used extensively to discuss project related issues among students and among students and teachers.

After the course was finished we provided students and all involved teachers with a simple evaluation form to evaluate the results of applying this tool in practice. Here are some of the highlights that we got from this evaluation.

First of all, there were no additional efforts on the teachers' side to prepare and conduct the course. The sample project and the alternatives for students had to be prepared anyway, regardless of the environment where the course was conducted. However, there was a need for a special lecture to explain students how to work with the tool. No other session with students were needed, because all the communication was going on in the online mode. This greatly reduced the time effort on the teachers' side because otherwise teachers would need to have 4-5 project meetings with students in the offline mode.

The evaluation of students' answers was quite positive as well. Firstly, they were asked if accomplishing a Web-based project was more difficult than accomplishing an offline project, which is a project with face-to-face project meetings. Since these students already had a number of projects in other university courses, which were accomplished in the offline mode, their answers might be seen as relevant. Only 5% of students answered that a Web-based project was more difficult to accomplish than similar projects that they had during their classes.

Secondly, they were asked if they see advantages in using communication and collaboration tools to work together on the project with other students. 80% saw such advantages and stated that the communication using the tool was in the most cases even better than in the offline mode, where the communication is usually restricted to the project meetings.

On the question if accomplishing such a Web-based project helped them to acquire additional skills, 90% students answered that they had acquired additional skills, and that there had been no negative difference between the skills acquired as compared with the more traditional projects. 85% of those 90% answered that they acquired these skills because Virtual Project Management Room provided an integrated environment needed to accomplish their task, e.g., they had communication with teachers and other students, possibility to discuss their results, to share their ideas with others, etc.

Finally, they were asked to assess the course and their overall assessment was 1.4, where 1 is the best possible mark on the scale from 1 to 5. The average assessment on the university is 2.5, and the average assessment on our institute is 2.

9.10 Conclusion and Future Work

The evaluation results clearly show that a didactics aware approach to implementing WBE systems and developing standards for WBE is a huge step in the right direction. However, some problems related to this approach need still to be resolved. For example, in order to support a new teaching scenario, e.g., a collaborative writing scenario, a new tool must be implemented. Obviously, each new scenario reflecting a particular didactic approach requires such a new tool. This, of course, can cost time and resources.

In our future work we plan to address this issue in the following way. Firstly, a modeling language for defining different didactic approaches should be developed. With such a language it should be possible to define all the components of a particular teaching scenario, such as educational material, user roles, the story of the scenario, student activities and others. For example, the story of a particular scenario might be defined as a number of learning actions that students need to accomplish. Student activities might include reading, writing, making tests, and others. Recently, some research efforts were undertaken trying to model didactics from different perspectives, such as constructivist perspective, activity-oriented perspective, etc. [22, 25]. We plan to investigate these research efforts and reuse as much results as possible coming from that research. Furthermore, the modeling language should be kept interoperable with the recent WBE standards. This will insure that standard compliant educational material can be easily incorporated within the system.

Secondly, a single generic tool capable of interpreting and executing teaching scenarios defined by means of the developed modeling language will be implemented into the WBT-Master. This tool will provide an integrated learning environment enclosing all educational material and other WBT-Master tools needed to support a particular scenario.

Thirdly, a number of typical teaching scenarios (such as scenarios presented in this chapter) should be modeled by the developed language and executed within the generic tool. We believe that different scenarios will share some common aspects. For example, communication in many different scenarios is usually based on a discussion forum and a chat room. Obviously, these two tools can be coupled together into a single communication component which may be reused as a module in different teaching scenarios. Thus, the modeling language must be component-oriented so that new teaching scenarios might be easily modeled by simply combining a number of already existing components.

Finally, we plan to implement a number of new teaching scenarios, such as collaborative writing, collaborative problem-solving, and others by combining and configuring existing and new components to meet the requirements of a particular teaching scenario.

References

1. Scorm - sequencing and navigation, version 1.3.1, http://www.adlnet.org/. Advanced Distributed Learning, http://www.adlnet.org/, 2004. Web link, http://www.adlnet.org/screens/shares/dsp_displayfile.cfm?fileid=998, last visited 2004-09-21.
2. Sharable content object reference model scorm, version 2004, 2nd edition. Advanced Distributed Learning, http://www.adlnet.org/, 2004. Web link, http://www.adlnet.org/index.cfm?fuseaction=DownFile&libid=648&bc=false, last visited 2004-09-21.
3. B. Barron, D. Schwartz, N. Vye, A. Moore, A. Petrosino, L. Zech, and J. Bransford. 1998 Doing with understanding: Lessons from research on problem- and projectbased learning. *Journal of the Learning Sciences*, 7:271–312.
4. Bb blackboard. Blackboard, 2004. Web link, http://www.blackboard.com/, last visited 2004-09-21.
5. R. Byrnes and A. Ellis. 2004 The distribution and features of learning management systems in australian universities and their role in student assessment. In *Proceedings of The Tenth Australian World Wide Web Conference*, Seaworld Nara Resort, Gold Coast, Australia.
6. Cic learning management systems (lms) survey. Committee on Institutional Cooperation, http://www.cic.uiuc.edu/, 2002. Web link, http://telr.osu.edu/surveys/cic-lms/reportFeb02.cfm, last visited 2004-09-21.
7. J. Dewey. 1938 *Experience and education.* Macmillan.
8. J. Dirkx and R. Lavin. 1991 Understanding and facilitating experience-based learning in adult education: The fourthought model. In *Proceedings of Midwest Researchto- Practice Conference.*
9. A. Hirumi. 2002 The design and sequencing of elearning interactions:a grounded approach. *International Journal on E-Learning*, 1(1):19–27.
10. D. L. Hobbs. 2002 A constructivist approach to web course design, a review of the literature. *International Journal on E-Learning*, 1(2):60–65.
11. Ims specifications, http://www.imsglobal.org/. IMS Global Learning Consortium, http://www.imsglobal.org/, 2004. Web link, http://www.imsglobal.org/specifications.cfm, last visited 2004-09-21.
12. D. Jonassen and D. Churchill. 2004 Is there a learning orientation in learning objects? *International Journal on E-Learning*, 3(2):32–41.
13. B. Jones, C. Rasmussen, and M. Moffit. 1997 *Real-life Problem Solving: A Collaborative Approach to Interdisciplinary Learning.* American Psychological Association, Washington DC, USA.
14. F. King and S. Puntambekar. 2003 Asynchronously conducted project-based learning: Partners with technology. *International Journal on E-Learning*, 2(2):46–54.
15. Ieee standard for learning object metadata, designation 1484.12.1-2002. Learning Technology Standards Committee, http://ltsc.ieee.org/, 2002. Web link, http://ltsc.ieee.org/wg12/par1484-12-1.html, last visited 2004-09-21.

16. A. Leasure, L. Davis, and S. Thievon. 2000 Comparison of student outcomes and preferences in a traditional vs. world wide web-based baccalaureate nursing research course. *The Journal of nursing education*, 39(4):149–154.
17. H. Mason, S. Rebelsky, and S. Luebke. 1999 Annotating the world-wide web. In *Proceedings of World Conference on Educational Multimedia, Hypermedia and Telecommunications*, pages 409–414, Norfolk, USA. Association for the Advancement of Computers in Education.
18. E. McKay, B. Garner, and T. Okamoto. 2002 Understanding the ontological requirements for collaborative web-based experiential learning. In *Proceedings of International Conference on Computers in Education (ICCE'02)*, page 356. IEEE.
19. D. Mioduser, R. Nachmias, O. Lahav, and A. Oren. 2000 Web-based learning environments: Current pedagogical and technological state. *Journal of Research on Technology in Education*, 33(1).
20. R. Oliver, B. Harper, T. Reeves, A. Strijker, and D. Westhuizen. 2002 Learning management systems: One size fits all? In *Proceedings of World Conference on Educational Multimedia, Hypermedia and Telecommunications*, pages 1498–1499, Norfolk, USA. Association for the Advancement of Computers in Education.
21. J. Oliviera, J. D'Ambra, A. Birk, J. Bund, M. Eberle, R. Ferreira, J. Hornung, P. Magana, and N. Scerbakov. 2001 Learning management systems and authoring tools: State-of-the-art. Technical Report IST-1999-11634, Fraunhofer IESE.
22. M. Packer and J. Goicoechea. 2000 Sociocultural and constructivist theories of learning: Ontology, not just epistemology. *Educational Psychologist*, 35(4):227–241.
23. M. Paulsen, D. Keegan, A. Dias, P. Dias, P. Pimenta, H. Fritsch, H. Fllmer, M. Micincova, and G. Olsen. 2002 Web-education systems in europe. Technical Report IST-1999-11634, Central Institute for Distance Education Research - FernUniversitt in Hagen.
24. E. Sorensen and E. Takle. 2001 Collaborative knowledge building in web-based learning: Assessing the quality of dialogue. In *Proceedings of World Conference on Educational Multimedia, Hypermedia and Telecommunications*, pages 1772–1777, Norfolk, USA. Association for the Advancement of Computers in Education.
25. A. Stutt and E. Motta. 2004 Semantic learning webs. *Journal of Interactive Media in Education*, 10.
26. J. Thomas. 2000 A review research on project-based learning. Technical report, The Autodesk Foundation.
27. J. Thomas, J. Mergendoller, and A. Michaelson. 1999 *Project-Based Learning: A Handbook for Middle and High School Teachers.* The Buck Institute for Education, Novato, USA.
28. S. Trapp. 2002 Coronet project - final report. Technical Report IST-1999-11634, Fraunhofer IESE.
29. M. Uljens. 1997 *School Didactics and Learning.* Psychology Press, Hove, East Sussex.
30. Webct learning without limits. WebCT, 2004. Web link, http://www.webct.com/, last visited 2004-09-21.

31. Owl web ontology language overview. World Wide Web Consortium W3C, http://www.w3.org/, 2004. Web link, http://www.w3.org/TR/owl-features/, last visited 2004-09-21.
32. Rdf vocabulary description language 1.0: Rdf schema. World Wide W Consortium W3C, http://www.w3.org/, 2004. Web link, http://www.w3.org/TR/rdfschema/, last visited 2004-09-21.

Index

S

T

U

V

W

MIX
Papier aus verantwortungsvollen Quellen
Paper from responsible sources
FSC® C105338

If you have any concerns about our products,
you can contact us on
ProductSafety@springernature.com

In case Publisher is established outside the EU,
the EU authorized representative is:
Springer Nature Customer Service Center GmbH
Europaplatz 3, 69115 Heidelberg, Germany

Printed by Libri Plureos GmbH
in Hamburg, Germany